Plant Form | An Illustrated Guide to Flowering Plant Morphology

Norantea guyanensis
A pitcher shaped leaf (bract 62) is associated with each
flower; see Figs. **88a**, **b** for early development.

Plant Form | An Illustrated Guide to Flowering Plant Morphology

Adrian D. Bell

School of Biological Sciences
University College of North Wales

With line drawings by
Alan Bryan

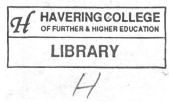
Oxford   New York   Tokyo
OXFORD UNIVERSITY PRESS
1991

Oxford University Press, Walton Street, Oxford OX2 6DP

Oxford New York Toronto
Delhi Bombay Calcutta Madras Karachi
Petaling Jaya Singapore Hong Kong Tokyo
Nairobi Dar es Salaam Cape Town
Melbourne Auckland

and associated companies in
Berlin Ibadan

Oxford is a trade mark of Oxford University Press

Published in the United States
by Oxford University Press, New York

British Library Cataloguing in Publication Data
Bell, Adrian D.
Plant form : an illustrated guide to flowering plant morphology.
1. Flowering plants. Morphology
I. Title
582.13041
ISBN 0-19-854279-8
ISBN 0-19-854219-4 pbk

Library of Congress Cataloging in Publication Data
Bell, Adrian D.
Plant form : an illustrated guide to flowering plant
morphology /
Adrian D. Bell : with line drawings by Alan Bryan.
1. Angiosperms—Morphology. 2. Botany—Morphology.
3. Angiosperms—Morphology—Atlases. 4. Botany—
Morphology—Atlases.
I. Bryan, Alan. II. Title.
QK641.B45 1990 582.13'044—dc20 90-34783
ISBN 0-19-854279-8
ISBN 0-19-854219-4 (pbk.)

Photoset by Cotswold Typesetting Ltd, Gloucester
Printed in Singapore by Times Printers Ltd

'The study of the external features of plants is in danger of being too much overshadowed by that of the internal features. The student, when placed before the bewildering variety of forms does not know where to begin or what to do to acquire information about the plants'.

WILLIS (1897)

'Horticulture is, undoubtedly, a great medium of civilization, and its pursuit is highly commendable, for it is impossible for anyone to study, even for a short period only, the structure, forms, and colours of plants, and benefits derived from the vegetable creation, without an elevation of thought, a refinement of taste, and an increased love of nature'.

B. S. WILLIAMS (1868)

'I have bought me a hawk and a hood, and bells and all, and lack nothing but a book to keep it by'.

BEN JONSON (1598)

Corypha utan
The single monopodial axis (section 250) finally terminating in an inflorescence after 44 years of growth. Model of Holttum (Fig. **291c**).

Preface

Flowering plants exhibit a fascinating array of external structures which can be studied with the naked eye or at most a simple hand lens. This is the science of plant morphology, the term being used here in the sense that excludes plant anatomy. Although an understanding of the form and external components of a plant should be the foundation of any botanical investigation, it is customary to rush ahead, delving deep into the plant, and thus either ignoring or missing the very features that the plant presents to the environment. The situation is very well-expressed by my namesake, Professor P. R. Bell (1985): 'In recent years the spectacular advances in molecular biology have generated such excitement that there has perhaps been a tendency for organisms to be overlooked. Biology must nevertheless remain ''organismic'', and the researcher who loses the concept of organisms seriously weakens his claim to be a biologist'. A blinkered attitude to plants probably commences at school level and continues through university. Excellent texts of plant morphology do exist, but they tend to presume a foundation of botanical education that is no longer available. The ground rules of plant morphology are, by and large, forgotten (Kaplan 1973a). The student of botany feels this defect but does not know how to resolve it; the academic conceals his ignorance. It is tempting to suggest that many an enthusiastic amateur horticulturist understands plants more intimately in terms of their morphology than does the average botanist. This criticism cannot be levelled, however, at the taxonomist who is armed with a great deal of morphological knowhow, heavily biased towards floral structure, and has at his disposal a profusion of terminology that is daunting to the beginner and expert alike. A guide is thus required for the benefit of both. This book is deliberately, I hope, attractive, the better to woo the budding botanist and the curious amateur plantsman. It is divided into two parts. The first part illustrates and explains much of the purely descriptive terminology involved in plant morphology, whilst the second part deals with an equally important but largely ignored aspect of morphology: constructional organization. The plant is developing, its organs are developing, most flowering plants branch, the branching patterns of the plant develop over time, and growth is dynamic. Cover of this aspect of plant morphology, which is of relevance to the ecologist and the population biologist (Harper 1980), culminates in an example drawn from the contemporary morphological world, that of the dynamic architecture of tropical trees. The author's fascination with plant morphology has been fostered by a providential succession of mentors, A. D. Prince at school, N. Woodhead at college, and P. B. Tomlinson ever since. Their teachings have one principle: if the morphology of a plant surprises you, then this is more likely to reflect your ignorance rather than an abnormality on the part of the plant. An unfortunate preoccupation with European plants in the past led morphologists to be taken aback by the exuberance of the world's vegetation, especially that of the tropics. But this is where the range of plant form can best be appreciated.

For this reason the plants illustrated in this book originate in all continents, and many will be unfamiliar to a reader confined to one geographical region. However, the same morphological features and details of constructional organization are repeated time and again in totally unrelated plants and the reader will recognize familiar forms if not familiar names.

In a sense this book can be treated as an illustrated dictionary to be consulted as necessary and in any sequence. With this in mind, the text and illustrations are extensively cross-referenced and the index annotated. The seasoned morphologist may be surprised to find equal space (one double-paged spread) allocated to such an insignificant feature as a stipel (58), and to such a vast topic as floral morphology (146) or the morphology of fruit and seed dispersal (160). Whole books have been devoted to these wider topics to which references are given, where appropriate, rather than the information being duplicated here. All the line drawings and diagrams are the work of Alan Bryan to whom I am clearly indebted. Alan's talent as an artist represents a happy combination of natural ability, an eye for detail, and a classical botanical training. Practically all the drawings and all the photographs have been taken from living plants, the exceptions being a few dried woody specimens, and a very few that have been adapted from existing illustrations. All the photographs were taken by the author (except 7 as noted) using an old Pentax Spotmatic II camera with a 105 mm lens (or occasionally a 35 mm lens) together with extension tubes where appropriate. Very frequently supplementary light was supplied by means of a pair of synchronized flash units mounted 15 cm to either side of the lens on a bar fixed to the camera body. Kodachrome 64 ASA was used throughout. I must thank many botanical gardens for allowing access to specimens: The Royal Botanic Garden, Edinburgh; The Botanic and Genetic Gardens of Oxford University; the Fairchild Tropical Garden, Miami, USA; The Botanic Gardens of Montpellier, France; the private gardens of M. Marnier Lapostolle, St. Jean Cap Ferrat, France; and the Treborth Botanic Gardens of the University College of North Wales, Bangor. Other photographic *sorties* have been made hither and thither and the author is grateful to those who have helped him visit various countries in Europe and particularly south and central America. I am very grateful to a great many people who have helped me in different ways to complete this guide: Nerys Owen for typing and most efficiently converting recorded tapes into word processor format without complaint and Josie Rodgers for taming the index with a suitable computer program; my colleagues at the School of Plant Biology, Bangor, for their encouragement and in particular Professor J. L. Harper for recognizing that plant morphology is a key subject. A number of kind people have commented on drafts of the manuscript at various stages. I am indebted to Professor F. Hallé of the Botanical Institute, U.S.T.L. Montpellier, to Professor P. Greig-Smith and Dr. and Mrs. N. Runham of Bangor, and to Professor P. B. Tomlinson of Harvard University. These good people have been able to point me in a better direction on a number of issues. Any errors that remain are my own; it is inevitable that some morphologists somewhere will take me to task on points of detail, rash generalizations, or personal opinions. I have relied on other people's plant identification in most instances and have followed the nomenclature of Willis (1973). Let me hide once again behind the axiom of my teacher at university; 'It is the plant that is always right'.

Contents

Root morphology

Stem morphology

Reproductive morphology

Seedling morphology

Vegetative multiplication

Grass morphology

Misfits

How to use this book

If you are already familiar with many aspects of plant morphology, you are likely to use this book in the manner of an illustrated dictionary, checking up on words and concepts but also allowing yourself to be side-tracked by the cross-referencing and thus discovering new aspects of plant form. If on the other hand the external features of flowering plants are a mystery to you, you should proceed to the introductory pages (4–16) and begin to learn your way about plants. At the same time thumb through the entries to gauge the scope of plant construction and interesting phenomena to watch out for. If a particular plant is presenting problems then you should proceed as directed in basic principles (4) unless you cannot resist the temptation to flick through the pictures first.

The book is divided into two sections. The first contains a descriptive account of the morphology of leaves, roots, stems, and reproductive parts. In the second section emphasis is placed on the way in which the organs of a plant are progressively accumulated. Thus, through both sections, a series of topics is presented usually with each being allocated one double-page spread. Each topic, such as petiole (40) is covered by a concise text, a representative photograph, and a selection of drawings or diagrams. This basic format is modified in some instances. All illustrations take their number from the page on which they occur. This layout allows comprehensive cross-referencing throughout, numbers in brackets (parentheses) in the text or in the figure legends sending the reader to either further examples of a given phenomenon or to the page on which an explanation for a particular term or concept can be found. Illustration numbers are in bold text; page numbers are not. The index is similarly comprehensive; entries of plant species names are qualified by their family and M or D for monocotyledon or dicotyledon (14) together with an annotation indicating the feature or features being described. Likewise topics are annotated for each entry to obviate unnecessary searching. In this manner the reader with a particular topic in mind will be able to locate its entry and also follow up associated features. The plants portrayed in the photographs and line drawings are identified by name (genus and species) and each drawing has an accompanying scale bar which represents 10 mm unless otherwise stated. Labels take the form of abbreviations which are listed at the foot of the page; again numbers in parentheses sending the reader to an explanation of a term or phenomenon, or to a supplementary illustration.

Reference is made to a limited number of detailed research works, and to monographs and books on specific topics. Other more general works concerned with the presentation of morphological information are Goebel (1900, 1905), Velenovsky (1907), Willis (editions up to 1960), Mabberley (1987), Troll (1935 to 1964), and Hallé *et al.* (1978, trees in particular). The phylogenetic approach is represented by Bierhorst (1971), Corner (1964), Eames (1961), Foster and Gifford (1959), Gifford and Foster (1989), and Sporne (1970, 1971, 1974). There is a huge vocabulary devoted to the description of flowering plants and which is principally to be found in the taxonomic literature. All floras will incorporate a glossary of terms, and a comprehensive lexicon is given in Radford *et al.* (1974). The morphological literature is also very extensive and it is not in the remit of this book to cover it here.

Introduction

Agave americana
Part of the inflorescence axis. Each main branch of this panicle (**141g**) is subtended by a bract (62), these forming a $\frac{3}{8}$ spiral phyllotaxis (**221b**).

Plant morphology is concerned with a study of the external features of plants. Literally it is the study of plant form. Anybody who has an interest in flowering plants must at some time have given a specimen more than a cursory glance and found various features that promote curiosity. There is a long history of interest in plant morphology. Perhaps the first scholars to become fascinated by the subject were Greek philosophers, especially Theophrastus (370–285 BC) who became bemused by plant form which he set about describing. He was concerned that an animal has a 'centre', a heart or a soul, whereas a plant has an apparently unorganized form which is constantly changing shape, i.e. it has no quiddity ('essence'). As plant morphology became a science, its purely descriptive role gained in importance and is still the first step in any taxonomic study. The 'pigeon-holing' aspect tended to lead to an inflexible facet of morphology (206) only recently shaken off. Nevertheless, morphology has undergone a number of transformations throughout history. Goethe (18th century) realized that a transition could be seen in the form of a leaf on a plant perhaps from foliage leaf to scale leaf to sepal and to petal. This is an example of the concept of homology about which much continues to be written (Tomlinson 1984; Sattler 1984). The foliage leaf on the plant and the petal, having the same developmental sequence and origins, are homologous structures. However, a foliage leaf and a flattened green stem (a cladode 126) are not homologous. They are merely performing the same activity and are analogous. Speculation

about homologous relationships forms the basis of phylogenetic studies, that is the identification of evolutionary sequences in plants. For a long time plant morphology became virtually submerged in this one field particularly after the advent of Darwin's theory of evolution. Another subject with which morphology is intimately associated is that of anatomy. All plant organs are made up of cells, and often the morphologist will want to study the development of an organ (its ontogeny) in order to understand its construction and affinities (e.g. 18). Developmental anatomy and details of vascular connections (the veins) within the growing plant are thus an essential feature of many morphological investigations. In some quarters morphology is indeed taken to cover both the external and the internal features of a plant. Dictionary definitions of 'morphology' may exclude any aspect of function in its meaning. However, it is very difficult to ignore the implication of function that is manifest in many plant structures; the function of a leaf tendril (68) in the vertical growth of a climbing plant cannot be disputed. This obvious utility leads to teleological statements ("the plant has evolved tendrils in order to climb") which should be avoided. Teleology is the philosophy that ascribes a deliberate intention of this nature to an organism. Plant morphology has always had a tendency to drift towards becoming a philosophical subject (e.g. Arber 1950), encouraging a contemplation of the inner meaning of the plant. In contrast the approach to be found here is hopefully more practical; the

intention is to provide an account of plant morphology as a working means of describing plant form, to emphasize that the knowledge of the development of a plant or plant part is as important as its final shape, and thus to stress that a plant is a growing dynamic structure in the light of which many morphological aspects of the plant should be considered.

Part I

Morphological description

Fig. 3. *Tragopogon pratensis*
Each fruit (an achene **157a**) borne on the capitulum (**141j**) has a pappus at its distal end aiding wind dispersal (cf. **155m**).

At first sight very many flowering plants have a familiar form. Each will have an underground branching root system which is continuous above ground with the shoot system (see Groff and Kaplan 1988, for a critical review of this classical axiom). The shoot system consists of stems bearing green (photosynthetic) leaves. The point on the stem at which one or more leaves are inserted (attached) is termed a node, and the interval of stem between two nodes is an internode (foresters use these terms in a different sense, a node being the position on a tree's trunk at which a whorl of branches is produced). In the axil of each leaf (i.e. in the angle between the upper side of a leaf and the stem) will be found a bud or the shoot into which the bud has developed. Such a bud is termed an axillary bud in distinction to one at the end of any shoot (a terminal bud). A leaf is said to subtend its axillary bud or shoot. There are a number of topographical terms that can be useful. The top of a leaf (or axillary shoot) is referred to as its adaxial surface, the underside being the abaxial surface. The part of a leaf (or shoot or root) furthest away from its point of attachment is the distal end of that organ. The end nearest the point of attachment is the proximal end. The various parts of such a conventional flowering plant are usually readily identified. A root will bear other roots and in some cases will also bear buds (roots buds, 178), but never leaves. A shoot will bear leaves with buds in their axils and may also bear roots (adventitious roots, 98). Leaves can drop off leaving scars (118) but there will still be a bud, or the shoot into which it has developed, in the axil of the leaf scar. Also the leaves may not be 'leaf-like' in appearance; they may be represented by insignificant scale leaves (64) or they may be modified in various ways, for example as spines (70, and see interpretation example 6) or tendrils (68). An underground structure, lacking leaves, is probably a root (94), but many plants have roots developing above ground and in some cases they are green (198). Conversely a great many plants (particularly monocotyledons 14) have underground stems (130) which most frequently will bear scale leaves and adventitious roots (98). It is important therefore to search a plant for clues as to the nature of its parts—roots bearing usually only roots, and stems bearing leaves of whatever morphology, each with its axillary bud. Leaves are usually relatively regular in their location (218), roots tend to be more irregular in location (96). It helps to remember that if the shoot is viewed with the youngest (distal) end uppermost (which is by no means necessarily the way it was growing) then a leaf will appear *beneath* each bud or shoot.

Many plants show simple or elaborate variations of these basic formats. The commonest variation is to find a shoot system that is sympodial rather than monopodial (250). An apparent departure from the leaf/axillary bud arrangement appears to take place at intervals on such an axis. (A relatively complicated example of sympodial growth is explained in sections 10 and 12.) Other factors to watch out for are leaves that lack associated axillary buds (244), leaves that subtend or apparently subtend more than one bud (236, 238), and buds that are not in the axils of leaves (but are located on roots, 178; in a displaced position on stems, 230; in a normal position but with leaves absent as occurs in many inflorescences, 140; or buds actually located *on* leaves, 74). Many plants do not have a resting stage and thus have no buds as such, only apical meristems (16, 262). A careful study of the plant will normally reveal such morphological features; in some cases a careful dissection of the youngest growing parts will help to identify the relationship of the parts and an understanding of development is useful (leaf 18, root 94, stem 112). It may be necessary to conduct a microscopical investigation of very early stages in some instances. Another factor that can at first sight mask the situation is that different organs can develop in unison and thus be joined together in the final form. This may be responsible both for the apparent displacement of buds (230), and the location of buds on leaves (74), as well as the fusion of parts (234). Again, there are structures to be found on stems and leaves that do not themselves represent leaves, or buds, or roots; they are termed emergences (76, 116) being developed from epidermal and subepidermal tissue. There are many ways in which the morphology of a plant can be recorded, a synopsis is given in section 8.

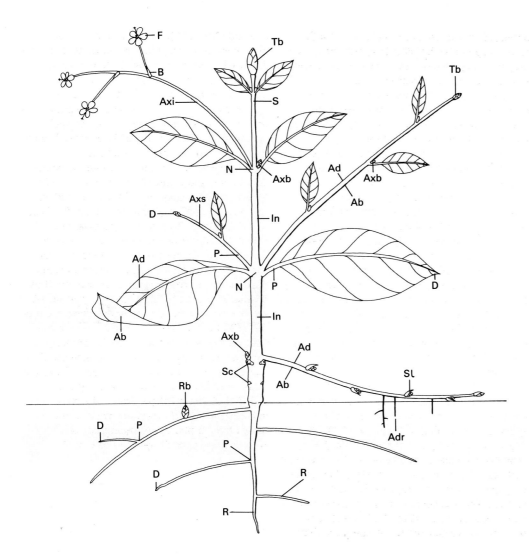

Fig. 5. Basic terminology. (There is no set convention for abbreviations used.) Ab: abaxial side. Ad: adaxial side. D: distal end. P: proximal end. Adr: adventitious root. Axb: axillary bud. Axi: axillary inflorescence. Axs: axillary shoot. B: bract. F: flower. In: internode. N: node. R: root. Rb: root bud. S: stem. Sc: scar. Sl: scale leaf. Tb: terminal bud.

A spine is usually identified as a tough, probably woody, structure with a sharp point. In morphological terms it may have developed from almost any part of the plant or represent a modification of any organ depending upon the species (76). In interpreting its nature, the basic 'rules' of plant morphology outlined in section 4 should be borne in mind. Is the spine subtended by a leaf, i.e. is it in the axil of an existing leaf or scar where a leaf has dropped off? If so, then the spine represents a modified stem (124). In many cases this will be confirmed by the presence of leaves (or leaf scars) on the spine itself (**125c**). However, a stem spine may be completely devoid of structures upon it. Also it may be encroached upon by the expanding stem on which it sits and all traces of its subtending leaf lost. It always pays therefore to look at a number of structures (spines in this example) of different ages as affinities are often more easily discovered in very young developing stages, perhaps even while still in a bud. For example it is possible that a spine which is apparently in the axil of a leaf actually represents one of the leaves of the axillary bud, and not the shoot axis itself (**203b**).

A spine that itself subtends a bud (or the shoot system into which the bud has developed of whatever form) will represent a modified leaf or part of a leaf (70). Nevertheless it may be very woody and very persistent, remaining on the plant indefinitely. Again, in order to discover exact origins, a developmental sequence should be studied. This may reveal that the spine or group of spines represents the whole leaf (**71a**) or perhaps just its petiole (**40b**), and then either the whole petiole or just a predictable part of it (**41c, d**). Frequently spines are found in pairs and then usually represent the stipules only of a leaf (**6, 57a, f**). If the spine does not appear to fall into the leaf/axillary bud format, then there are still a number of possible explanations. The spines of the Cactaceae for example (202) in fact represent modified leaves, but this is not at all apparent from casual observation; some plants have morphologies that are only decipherable with detailed, usually microscopic, study. A spine may be formed from a root (**107d**). It will therefore not be associated with a leaf, and its root origins will have to be identified by section cutting to show that it is endogenous (having its origin deep in existing tissue) in development and that it has a root anatomy and probably a root cap in its early stages (94). Leaf and stem spines are exogenous, arising at the surface of their parent organ (18, 112). Leaf, stem, and root spines will contain veins (vascular bundles). A fourth category of spine may be encountered and may be found on a leaf (76) or stem (116). This type of spine (or 'prickle') is not in the axil of a leaf, does not subtend a bud, and is not endogenous in origin and also lacks vascular tissue. It is termed an emergence, and develops from epidermal and just subepidermal tissue. Emergences are usually much more easily detachable than stem, leaf, or root spines. This general account of 'spine' could equally well have taken 'tendril' (68, 122) as its theme. The same 'rules' apply, bearing in mind that some plants are nonconformists (206) and that displacement and merging of different organs can take place (bud displacement 230, adventitious buds 232, adnation 234, teratology 270).

Fig. 6. *Acacia sphaerocephala*
Persistent stipular spines (56).

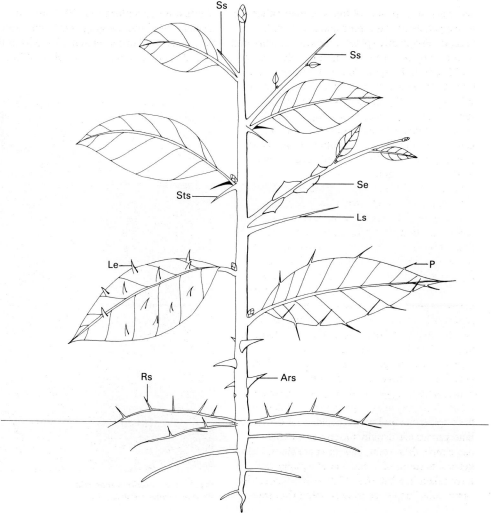

Fig. 7. The range of structures forming spines (see also bract spine **63b**, inflorescence spine **145d**). Ars: adventitious root spine (106). Le: leaf emergence (76). Ls: leaf spine (70). P: prickle (76). Se: stem emergence (116). Ss: stem spine (124). Sts: stipular spine (56). Rs: root spine (106).

8 | Methods of description

We have in our library a book entitled *Natural Illustrations of the British Grasses* edited by F. Hanham (1846). The preface tells us that the success of the book depends entirely upon the illustrations which are, in fact, dried specimens of actual grasses of which 62 000 were collected for the edition. Actually, plants lose many morphological features when pressed and dried, such as colour, hairiness, and three-dimensionality [Corner (1946) points out that many plants with spiral phyllotaxy (218) have been recorded as having distichous leaf arrangement when studied from pressed specimens alone]. Nowadays it is customary to record aspects of plant morphology using a combination of illustrative methods. A photograph is an obvious choice, colour being preferable to black and white as the brain can have trouble deciphering grey tones. However, a photograph alone is not enough as it is likely to contain a great deal of distracting 'noise' both on the subject itself and in the background. It is better to augment or replace a photograph with line drawings (see for example **106** and **205b, 62b** and **63b**). These will range from accurate detailed representations of actual specimens to drawings in which line work is kept to a minimum in the interest of clarity (but not at the expense of accuracy), to various diagrammatic versions of the actual specimen or even hypothetical diagrams (such as **183, 253**) of its construction (an example of these various possibilities is given for one specimen on pages 11 and 13). A particularly useful feature of a greatly simplified diagram is that it can take the form of a cartoon series to illustrate the developmental sequence of a particular morphological feature (e.g. **11b, c, d, e**). The simplest form of this type of illustration can be termed a 'stick' diagram in which the thickness of the organ is ignored and a stem, for example, can be represented as a fine line with leaves and axillary shoots portrayed symbolically (**9b**). However, stick diagrams suffer the same limitations as pressed plants in that it is difficult to retain three-dimensionality. It is for this reason that the combination of photograph and/or accurate line drawing plus diagrammatic portrayal is most informative. In addition the relative juxtaposition of parts can be indicated by another type of diagram, the ground plan (or specifically floral diagram for flowers and inflorescences 150). The ground plan depicts a shoot system, or flower, as if viewed from directly above. The position of leaves, buds, and axillary shoots are located on the plan in their correct radial (i.e. azimuth) positions, the youngest,

Fig. 8. *Euphorbia peplus*
The inflorescence consists of a symmetrical set of cyathia (144, **151f**), each resembling a flower but in fact composed of numerous much reduced flowers.

distal organs are sited at the centre of the diagram and are best drawn in first, and the oldest (lowest, proximal) organs are at the periphery. All components are represented in a schematic fashion, there being a more or less conventional symbol for a leaf, for example. Two leaves which are in reality the same size, may appear as different sizes in the plan, the more distal, inner, being drawn the smaller. Conversely, a proximal, small, scale leaf may appear on the plan as a larger symbol than a more distal foliage leaf. Nevertheless the ground plan approach is a most valuable adjunct to the other forms of diagrammatic illustration as it reveals underlying patterns of symmetry (228) or their absence, in the construction of shoot systems (including flowers). The four types of monochasial cymes illustrated in 'stick' form (141s, t, u, v) are repeated here in ground plan form (9d, e, f, g). The power of illustration in conveying detailed information should not be belittled: "Artistic expression offers a mode of translation of sense data into thought, without subjecting them to the narrowing influence of an inadequate verbal framework; the verb 'to illustrate' retains, in this sense, something of its ancient meaning—'to illuminate'" (Arber 1954).

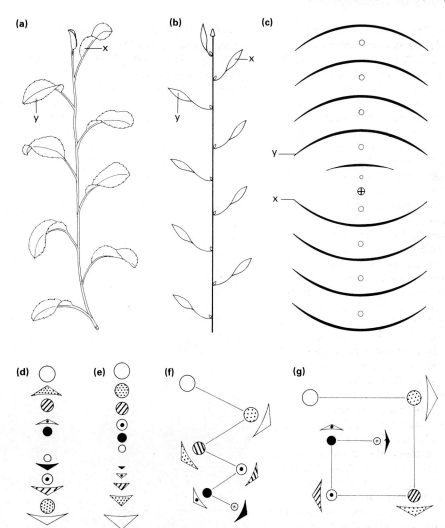

Fig. 9. a) Shoot drawn from life. b) 'Stick' diagram of 'a' indicating its component parts. c) Plan diagram of 'a' showing relative locations of parts. The same two leaves are labelled x and y in each case. d)–g) Plan diagrams of the four types of monochasial cyme shown in stick form in Fig. 141s–v. Each shoot symbol (a circle) has the same patterning as that of the leaf symbol in whose axil it is located.

Many species of *Philodendron* (Araceae) have a distinct morphology which presents a number of the features described in this volume. The shoot organization is not immediately apparent from casual observation, but can be interpreted by study of the plant as it develops (Ray 1987a, b). A description of this admittedly complex plant is given here in order to demonstrate the use of various descriptive methods. Figure 10 shows the general features of a young vegetative plant of *Philodendron pedatum* collected in French Guiana. The photograph gives an overall impression of the plant, but this is enhanced by the accompanying line drawing (11a) which eliminates confusing detail and background and allows the major features of the plant to be labelled. At first sight the stem of the plant appears to bear an alternating sequence of large scale leaves, represented by their scars except for the youngest, and foliage leaves. Close scrutiny will fail to find an axillary bud associated with the foliage leaf and it will be noticed that two leaves appear on one side of the stem followed by two more or less on the opposite side as indicated in Fig. 13e. If the shoot represents a monopodial system (250) then the plant must have an unusual phyllotaxis. Figure 13d illustrates this in a simplified manner and this is repeated in more simple 'stick' fashion in 13e. The adventitious roots (98) present at each node (11a) are omitted from these diagrams for simplicity. Close study of *Philodendron* shows that the shoot axis is in fact sympodial (250) in its construction, each sympodial unit (shown alternately hatched and unhatched in Fig. 13b) terminates as an aborted

apex (244) in this juvenile plant (in a mature plant these distal ends of sympodial units could be represented by inflorescences). The aborted apex is usually barely visible. In a sympodial sequence, growth of the shoot is continued by the development of a bud on the previous sympodial unit. In the case of this *Philodendron* species, the bud that develops is one of two buds (accessory buds 236) in the axil of the first leaf of each sympodial unit (i.e. the prophyll 66), with the second leaf, the foliage leaf, subtending no bud.

(Continued on page 12.)

Fig. 10. *Philodendron pedatum*
Young plant. Latest developing leaf is emerging from the protection of a prophyll (66).

(a)

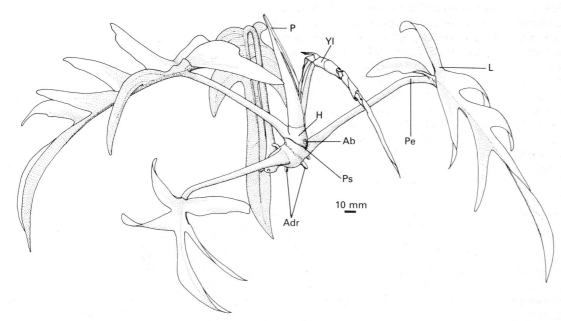

10 mm

Fig. 11. a) *Philodendron pedatum* Young plant (cf. **10**). b)–e) Developmental sequence indicating how the upper bud (hatched) of the pair in the axil of a prophyll displaces the shoot apex to become the next sympodial unit. Compare with Fig. **13b**. Ab: accessory bud. Adv: adventitious root. H: hypopodium. L': leaf lamina. P: prophyll. Pe: petiole. Ps; prophyll scar. Sa: shoot apex. Yl: young leaf.

(b)   **(c)**   **(d)**   **(e)**

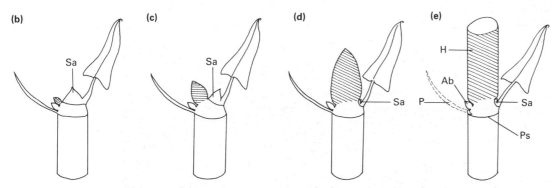

Fig. 12. *Philodendron pedatum*
Close up of stem showing prophyll (66) scars and
adventitious roots (98). Shoot apex bent over to upper right.
This photograph is represented by the upper half of Fig. **13a**.

The developmental sequence of events in
Philodendron (page 10) is shown in
Fig. **11b, c, d, e**. The upper, hatched, bud in the
axil of the prophyll develops rapidly displacing
the shoot apex to one side and leaving behind the
second, lower bud in the axil of the prophyll.
This bud can be seen in Fig. **12**. The sympodial
nature of the axis is represented in Fig. **13a, b, c**.
Figure **13a** gives a stylized appearance of the
shoot, **13b** indicating the locations of the
successive sympodial units, and **13c** (a stick
diagram) giving a truly diagrammatic
representation of the shoot construction. Stick
diagrams such as this are extremely useful in
conveying plant construction with a minimum of
background noise. However, it is not easy to
illustrate the three-dimensional aspect of the
morphology by this means; the relation of one
leaf to another can be portrayed in the form of a
ground plan diagram (see page 8), being the
vegetative equivalent to a floral diagram (150).
Figure **13f** indicates the juxtaposition of parts of
the sympodial specimen (**13c**) and **13g** the plan
that would be found if the shoot was monopodial
(**13e**). The precocious elongation of each
sympodial unit results in a substantial bare
length of stem between the point of attachment
of this side shoot on its parent, and the node
bearing the propyll ('H' in Fig. **13a**). Such a
portion of stem, proximal (8) to the first leaf on a
shoot is termed a hypopodium (see syllepsis 262).
So in this young state, this particular species of
Philodendron has a sympodial shoot system, each
unit of which bears just two leaves and two buds
(both in the axil of the first leaf). However, it is

recognized that plants can go through a sequence
of morphological forms as they develop, each of
which can be described as an age state (314).
The seedling *Philodendron* probably undergoes a
period of monopodial establishment growth (168)
before switching to the sympodial sequence
described here; this is deduced from the activity
of the second prophyll bud on this plant which, if
it develops, does so monopodially at first (and
thus represents a reiteration 298). Furthermore
the sympodial sequence shown, with prophyll
scale leaf followed by a single foliage leaf, itself
gives way to a different sympodial sequence in
which the second leaf aborts when about 1 cm in
length, whilst the hypopodium is greatly
extended (**66a**). In this state the plant climbs
rapidly. It has yet to reach a mature stage with
enlarged foliage leaves and a reproductive
capacity. Details of the very precise sympodial
sequences found in the family Araceae are given
by Ray (1988).

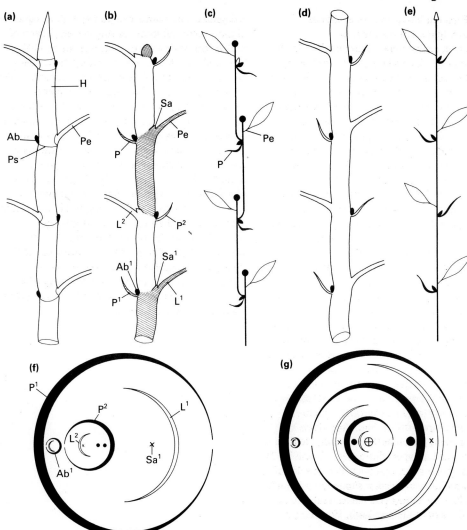

Fig. 13. Shoot construction of *Philodendron pedatum* (cf. **11**). a)–c) Alternative methods of depicting the sympodial sequence. d), e) Diagrams showing the superficial monopodial appearance. f) Plan view of 'b'. g) Plan view of 'd'. Ab: accessory bud. H: hypopodium. L: foliage leaf. P: prophyll. Pe: petiole. Ps: prophyll scar. Sa: shoot apex. The specific features labelled in the lower half of b) (Ab^1, L^1, L^2, P^1, P^2, Sa^1) have corresponding labels in the ground plan f).

This book describes the morphology of flowering plants. Taxonomically these represent seed producing plants (Spermatophyta) as opposed to spore producing plants, and further represent plants in which the seeds are contained within a fruit (154) (Angiospermae, or Angiosperms, from the Greek 'angeion'—a vessel, i.e. container, +sperma, seed) in distinction to the Gymnospermae in which the seeds are naked (e.g. principally conifers). The flowering plants fall naturally into two categories: the dicotyledons and the monocotyledons (D or M in the index). The differences between these two groups are marked and the botanist can usually tell the one from the other at a glance even when meeting a plant for the first time. There are however, also plenty of species whose monocotyledonous or dicotyledonous affinities are not at first sight apparent. Monocotyledons include palms, gingers, lilies, orchids, grasses, sedges, bananas, bromeliads, and aroids; dicotyledonous plants include most trees and shrubs, and herbaceous and woody perennials.

As their name suggests, monocotyledons have one cotyledon (seedling leaf 162) whereas dicotyledons almost invariably have two. The flowers of monocotyledons usually have components in sets of three whereas dicotyledons very rarely have flower parts in sets of three; four or five being more typical. The principal difference between the two groups of a morphological nature involves their mode of growth. The stems of monocotyledons, with few exceptions, lack the ability to increase continuously in girth, i.e. they lack a meristematic cambium (16). Very many dicotyledons do possess this tissue and their stems and roots can grow in diameter keeping pace with the increase in height. Increase in stature in a monocotyledon takes place by means of establishment growth (168) in which each successive internode (4) or sympodial component (250) is wider than the last. One consequence of this difference and of a difference in vascular anatomy is that the leaf of a monocotyledon is usually attached more or less completely around the stem circumference at a node, whereas a dicotyledon leaf is more often attached on one relatively narrow sector of the stem circumference. More fundamentally, the first root (radicle 162) of a dicotyledon is quite likely to increase in size as the plant above grows. This also means that the proximal (4) end of the root system increases in size as more and more lateral roots develop distally; no bottleneck or mechanical constriction is formed. This cannot occur in the majority of monocotyledons and the roots initially developing from the embryo are soon inadequate in diameter to serve the growing plant. All monocotyledons develop an adventitious root system (98), i.e. numerous additional, but relatively small roots extend from the stem of the plant. This is particularly well seen in rhizomatous (130) or stoloniferous (132) monocotyledons which are usually sympodial (250) in construction and in which each new sympodial unit will have its own complement of new adventitious roots. The lack of a cambium in monocotyledons is also reflected in the limitation of their above-ground branching. When a bud of a monocotyledon develops into a shoot it must usually do so by the progressive increase in diameter of each successive internode as in the establishment of a seedling (**169c**); the whole base of the new plant or branch cannot grow in diameter as is possible in dicotyledons. The consequences for aerial branching would be a mechanically unstable constriction at the point of attachment of the branch. Monocotyledons that do branch aerially either have very slender branches (e.g. bamboos 192) or branches supported by prop roots (**100**) or gain support by climbing (**98**) or form a mechanically sound joint by precocious enlargement of the side branch at the time when the parent stem is itself still growing, i.e. the two develop in unison (**11b, c, d, e**). A commonly stated 'rule of thumb' to distinguish a dicotyledon from a monocotyledon is that a dicotyledon leaf probably has a petiole (40) and is reticulately (net) veined (34) whereas a monocotyledon leaf usually lacks a petiole and is parallel veined. However, there are innumerable exceptions to both these sets of generalizations (e.g. **21b, 35**).

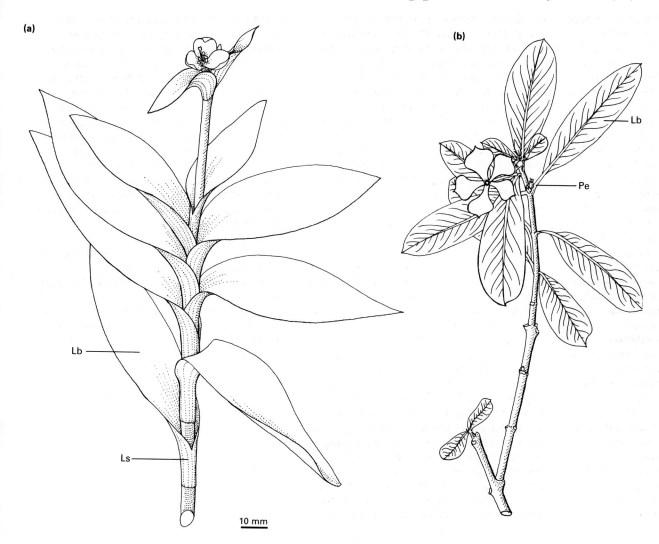

(a)

(b)

Lb

Pe

Lb

Ls

10 mm

Fig. 15. a) *Setcreasea purpurea* (Commelinaceae), a monocotyledon. b) *Catharanthus roseus* (Apocynaceae), a dicotyledon. Lb: leaf blade. Ls: leaf sheath. Pe: petiole.

Fig. 16. *Cyclamen* cv
A developmental 'mistake' (270) in which a flower bud is joined to the stem axis (adnation 234) and is being pulled away from its subtending leaf so much that the flower stalk (pedicel 146) has snapped.

All the various organs and morphological features of a plant are made up of cells, growth and development taking place in localized regions of active cell division and enlargement. Such regions are termed meristems (18, 94, 112) and typically there is a meristematic zone at the apex of every shoot (a shoot apical meristem) and every root (a root apical meristem) on the plant. The apical meristem of a shoot may be protected, particularly if in a resting stage, by older tissues and organs such as scale leaves to form a bud (264). However the shoot apical meristems of many plants undergo more or less continuous growth and do not rest in bud form, thus these axillary shoot apical meristems develop contemporaneously with that of the supporting axis (syllepsis 262). Lateral roots (96) and adventitious roots (98) develop from root apical meristems that arise deep in the tissue of existing roots or stems, respectively (94). Meristematic activity at shoot apices gives rise to new leaves. The first stage is the appearance of a leaf primordium (18), in which cell division of specific meristems results in a leaf shape characteristic for the plant. The leaf edge expands as a result of meristematic activity of the marginal and plate meristems of the leaf, for example (19c, d). In the majority of monocotyledons (14) and in many dicotyledons, the entire plant body is built up by cell division and enlargement at the apical meristem of shoot and root. This is referred to as the primary plant body. In numerous dicotyledons, a second form of meristematic activity can also take place which results in the enlargement of the existing primary plant body.

Within the primary stem and root there is a cylinder of cells that retains its meristematic properties. This zone, the cambium, is sometimes referred to as a lateral meristem to distinguish it from an apical meristem. Cell division within the cambial cylinder leads to expansion in girth of the stem or root by the production of secondary tissue including vascular tissue. A constantly enlarging plant, such as a tree, is built up in this way. Reference should be made to an anatomical textbook (e.g. Esau 1953; Cutter 1971) for a comprehensive account. A few monocotyledons have a similar process producing secondary tissue by means of lateral meristem activity and form branched tree structures (e.g. *Cordyline, Dracaena*). Other monocotyledonous trees, such as palms, gain their stature following establishment growth (169c). A second type of lateral meristem, again in the form of a cylinder, may be present just beneath the surface of a stem or root; this is termed a phellogen, or cork cambium, and gives rise to the bulk of the bark (114).

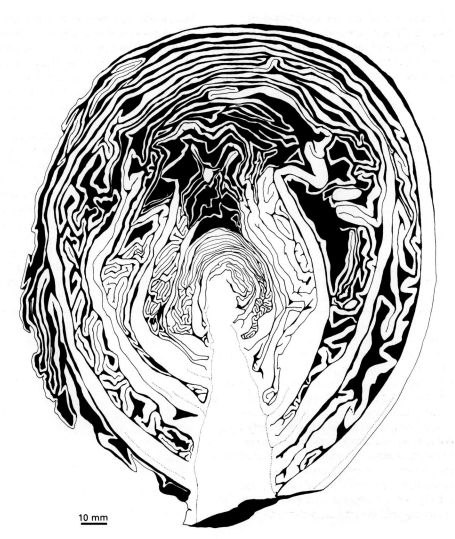

10 mm

Fig. 17. *Brassica oleracea*. Longitudinal section through the shoot apex. The leaves show crumpled vernation (**149c**).

Fig. 18. _Plumeria rubra_
Shoot apex before flowering showing sequence of leaf development.

New leaves develop at the surface of the apical meristem of a shoot which is itself extending by cell division and enlargement (16). Thus each new leaf, termed a leaf primordium, is left behind on the flanks of the axis as the shoot extends. The most recent leaf primordium to appear at the apical meristem is the least developed, and successively older leaf primordia are progressively more elaborate due to the activities of meristems within the leaf itself. The primordium of a dicotyledonous leaf is usually confined to a relatively narrow sector of the shoot circumference whereas in contrast a monocotyledonous leaf primordium is initiated and therefore develops around most, if not all, of the shoot apex. Thus very young dicotyledonous leaves are peg-like structures (**19a**) and correspondingly young monocotyledonous leaves are collar-like structures (**19b**) surrounding or even arching over the shoot apex. The sequence in which new primordia appear at the apex will give the plant its particular phyllotactic arrangement (218). A leaf primordium will continue to grow in size and gradually attain its destined determinate size and shape. Increase in leaf size results from an increase in cell numbers followed by an increase in cell size. Cell division is loosely confined to identifiable meristematic regions (16) in the leaf and it is the differential activity of these regions that produces different leaf shapes. At first the apical meristem of the leaf is active and the leaf elongates, subsequently leaf elongation results from activity of the intercalary meristem (**19c**). This meristem can have a prolonged activity, in grasses for example (180).

A horizontally flattened shape (bifacial or dorsiventral) will result if the marginal meristems become active (**19c**), leaf width being increased by division in the plate meristems (**19d**). If the marginal meristem is only active at sites dispersed along the leaf edge then a pinnate leaf (22) will result. Each leaflet of a pinnate leaf develops from an isolated patch of marginal meristem and will be organized in a similar manner to a whole simple leaf (**19e**). The midrib becomes thicker than the lamina due to cell division of the adaxial meristem (**19d**). If the adaxial meristem continues to contribute to thickness in this region and at the same time the marginal meristems are inactive, then the leaf will be flat in the vertical plane (lateral flattening) and result in an ensiform leaf (86). Between monocotyledons and dicotyledons there is a fundamentally different emphasis of meristematic activity towards either the base of the very young leaf primordium (lower leaf zone) or the apex of the primordium (upper leaf zone) (20). Also in some instances controlled cell death plays a part. This is responsible for the indentations and holes that appear early in the development of leaves of some members of the family Araceae (10) and occurs in the formation of compound leaves in the palms (92). Areas with meristematic potential may remain on parts of a leaf and subsequently develop into vegetative (**233**) or inflorescence (**75g**) buds. In a few plants the apical meristem of the leaf remains active and the leaf can continue to grow apically for an extended period (90).

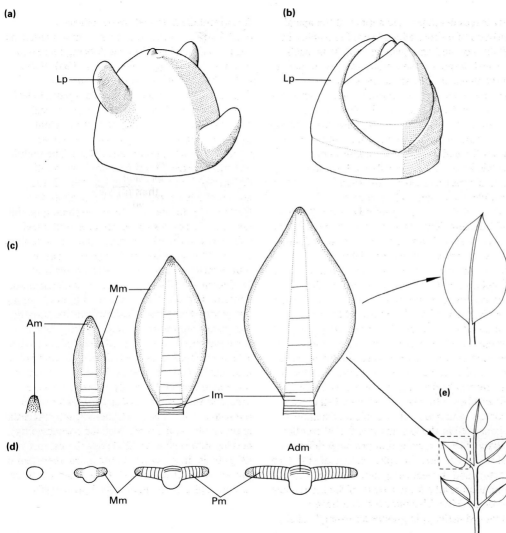

(a)

(b)

Lp

Lp

Fig. 19. a) Diagrammatic representation of the shoot apex of a dicotyledon, and b) of a monocotyledon. c) The meristematic zones of a simple developing leaf seen from above, and d) in section. e) The same components apply to the leaflet of a compound leaf. Adm: adaxial meristem. Am: apical meristem (of the leaf). Im: intercalary meristem. Lp: leaf primordium. Mm: marginal meristem. Pm: plate meristem.

(c)

Mm

Am

Im

(e)

(d)

Adm

Mm

Pm

Studies of the very early sequences of growth of leaf primordia indicate that the two ends of the primordium, the distal (apical) end and the proximal (basal) end, give rise to specific parts of the mature leaf (Kaplan 1973*b*). A fundamental difference is found in the development of monocotyledonous and dicotyledonous leaves. In many 'typical' dicotyledonous leaves the proximal end of the primordium (lower leaf zone) will develop into the leaf base which may or may not ensheath the stem (50) together with the stipules if present (52). The distal end of the primordium (upper leaf zone) develops into the dorsiventrally flattened leaf blade (**21c**) (lamina) or laterally flattened phyllode in the case of some *Acacia* species (44). Subsequent activity of an intercalary meristem and an adaxial meristem (18) may separate the base from the lamina by the development of a unifacial (i.e. more or less cylindrical) petiole (40). However, if the relative development of the lower and upper leaf zones of a dorsiventral monocotyledonous leaf are monitored, it is found that the whole of the leaf, sheathing base plus lamina (**21e**) and also the petiole if present (**21b**), is derived from the lower leaf zone. The upper leaf zone hardly contributes to the mature leaf structure at all but may be present in the form of a unifacial rudimentary 'precursor tip' at the apex of the leaf (**20a, b**). Some monocotyledons have unifacial leaves (86), the distal unifacial portion being substantially longer than the basal sheath (**21a**). Studies of development of these leaves show that the unifacial portion develops from the upper leaf zone and is equivalent to the precursor tip of

bifacial monocotylendonous leaves and thus also in these cases equivalent to the lamina of a dicotyledonous leaf. Indeed the development of such a unifacial monocotyledonous leaf is virtually identical to the development of the unifacial leaf that occurs in some dicotyledons although the latter may show rudimentary

pinnae (e.g. **21d**). Conversely a few dicotyledon leaves are equivalent in their development to monocotyledon leaves in that the bifacial portion develops from the lower leaf zone (**89c**). In a heteroblastic sequence (**29d**) of leaves the change of leaf shape emphasizes changes in relative activity of upper or lower leaf zones.

Fig. 20a. *Musa* sp.
Precursor tip at distal end of unrolling leaf.

Fig. 20b. *Sansevieria* sp.
Terete (86) precursor tip at distal end of leaf.

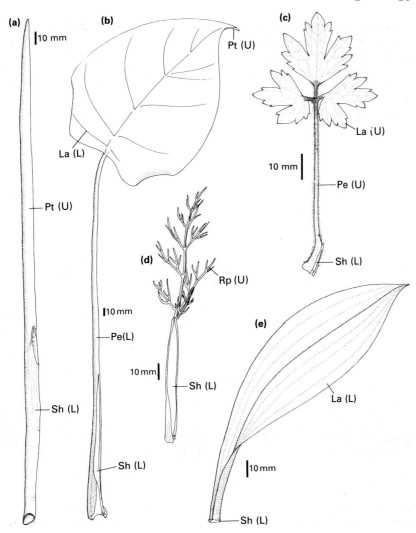

Fig. 21. Comparison of single leaves of monocotyledons (M) and dicotyledons (D). a) *Allium crepa* (M), b) *Monstera deliciosa* (M), c) *Ranunculus repens* (D), d) *Foeniculum vulgare* (D), e) *Rossioglossum grande* (M). La: lamina. Pe: petiole. Pt: precursor tip. Rp: rudimentary pinnae. Sh: sheath. U: upper leaf zone. L: lower leaf zone.

Fig. 22. *Calathea makoyana*
Translucent simple leaf with chlorophyll confined to specific areas mimicking a pinnate leaf.

The shape of a leaf depends on its development, usually in terms of cell division and enlargement, but also due to cell death in some cases (18). There is a very precise and extensive terminology applied to the lamina shapes of simple (**22, 23a, c, f**) leaves and to the individual leaflets, each of which has its own stalk, of compound leaves (**23b, d, e, g**). Such terms refer to the lamina base, tip, the margin, and overall geometry. Thus a simple leaf may be described as widely ovate, apex caudate (with a tail) and base cordate (**35b**), i.e. heart-shaped with a long drip-tip. Definitions of these terms in common usage may be found in the glossary of any flora. A compound leaf may be simply pinnate, leaflets or pinnae being arranged in an alternate (**124a**) or opposite (**27b**) fashion, sometimes the one merging into the other (**23b**). There may be a single terminal leaflet (imparipinnate **57f**) or this may be absent (paripinnate **27a, 23e**) or represented by a pointlet (**79**). If the leaflets are of variable size the compound leaf is described as interruptedly pinnate (**271h**). The central stalk bearing the leaflets is termed the rachis. In a bipinnate leaf the rachis bears rachilla on which the leaflets (pinnules) themselves are inserted (**23e**). The stalk of each individual leaflet can be termed a petiolule. If all the petiolules are attached at one point the compound leaf is palmate (**27e**). More precise terms can be applied to palmate leaves having a consistent number of leaflets, e.g. bifoliate and trifoliate (**23g**). The term unifoliate can be applied to a simple leaf in which the lamina is articulated on the petiole (**49d**). (Continued on page 24.)

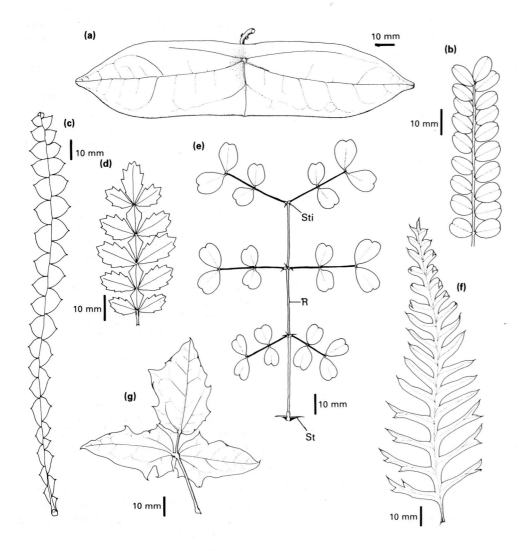

Fig. 23. Shapes of single leaves. a) *Passiflora coriacea*, simple; b) *Sophora macrocarpa*, simply pinnate; c) *Banksia speciosa*, pinnatisect, lobed to the mid-rib; d) *Weinmannia trichosperma*, simply pinnate; e) *Rhynchosia clarkii*, bipinnate; f) *Grevillea bougala*, pinnatifid, lobed nearly to mid-rib; g) *Lardizabala inermis*, trifoliate. R: rachis. St: stipule (52). Sti: stipel (58).

More elaborate configurations exist and will be described in specialist works on taxonomy and systematics (e.g. Radford *et al.*, 1974). All these terms apply to dorsiventrally flattened leaves. In addition leaves may be variously three-dimensional (24), laterally flattened or radially symmetrical (86), may bear various structures on their surface (74, 76, 78, 80), or may be represented by tendrils, hooks (68), and spines (70).

Plant species exhibit a remarkable range of leaf shapes (22). Indeed the manner of development of a leaf (18) permits almost any configuration, subject to mechanical constraints. By no means all leaves are bilateral (86, 88), and many have a three-dimensional construction. If only an occasional leaf on a plant is a bizarre shape, it is likely to represent an example of teratology (270) or possibly gall formation (278). Leaves may bear other organs—epiphylly (74). The illustrations here (**24, 25**) depict just a very few of the leaf shapes that can be found; many other examples could be used. The base of the leaf and its juxtaposition with the stem (sheath 50) also presents a range of forms, again often of a three-dimensional construction. The petiole of the leaf (40) and the mid rib (rachis) of a compound leaf (22) may be winged, the manner in which the wings meet the stem varying considerably. Four of the more usual forms are auricled (**25c**), amplexicaul (**29c**), perfoliate (**25b**), and decurrent (**24**). The leaves on any one plant may have a range of shapes either of distinctly different forms (30) or in developmental series (28).

Fig. 24. *Onopordum acanthium*
The decurrent leaf bases extend down the stem as wings.

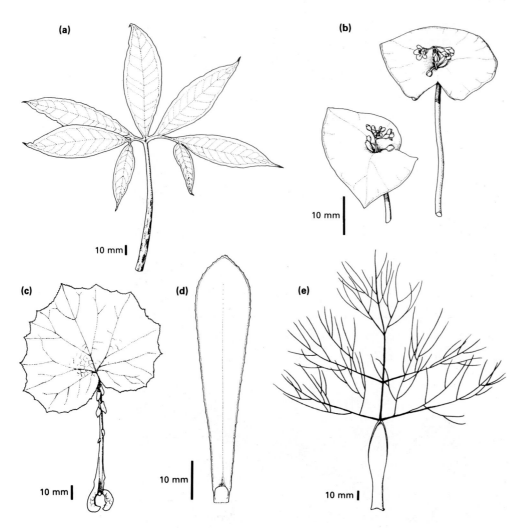

Fig. 25. Shapes of single leaves. a) *Sauromatum guttulatum*, palmate; b) *Montia perfoliata*, perfoliate; c) *Senecio webbia*, auricled at petiole base; d) *Othonnopsis cheirifolia*, simple; e) *Foeniculum vulgare*, multi-pinnate.

(a)

(b)

(c)

(d)

(e)

10 mm

10 mm

10 mm

10 mm

10 mm

Leaves of all shapes vary very much in their degree of symmetry. Asymmetry is much more pronounced in some species than others and may occur to varying extents on the same plant. Asymmetry in a given species can be precise and repeated by all leaves (**243**) or may be imprecise so that each leaf has a unique shape in detail (**27d**). Thus simple leaves are frequently asymmetrical at their base and then the shoot as a whole may or may not be symmetrical due to mirror imagery of leaves to left and right (**32**). In the Marantaceae, leaves are more or less asymmetrical about the midrib (**22**), the wider more convex side being rolled within the narrower straighter side in the young state. The wide side may be to the left or right viewed from above and this may or may not be consistent in a given plant or species. The arrangement can be antitropous (**27h**) or more frequently homotropous (**27f**). A theoretically possible alternative homotropous configuration (**27g**) does not seem to occur (Tomlinson 1961). Pinnate leaves frequently show a degree of asymmetry both in the apparent absence of some pinnae (**45**) and in the admixture of first order leaflets with second order rachillae (**47c**). Compound leaves with symmetrically opposite leaflets at their proximal ends may have asymmetrically alternate leaflets at their distal ends (**69f**) or vice versa (**271h**).

Fig. 26. *Manihot utilissima*
Palmate leaf.

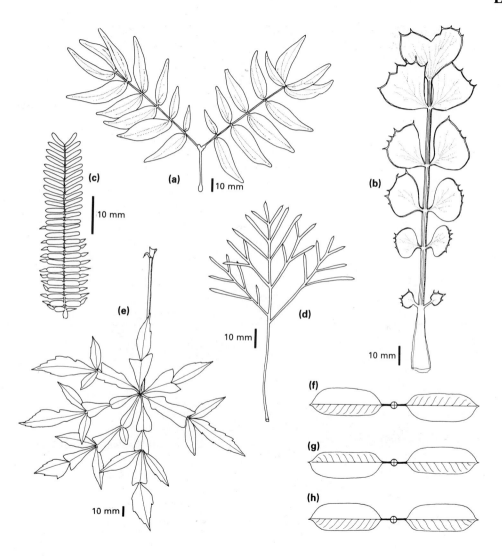

(c)

(a)

| 10 mm

10 mm

(e)

(d)

10 mm |

(b)

10 mm |

(f)

(g)

(h)

10 mm |

Fig. 27. Shapes of single leaves. a) *Calliandra haematocephala*, bipinnate; b) *Azilia eryngioides*, pinnate; c) *Acacia hindisii*, single pinna cf. **79**; d) *Isopogon dawsonii*, bipinnate; e) *Cussonia spicata*, palmate; f)–h) Asymmetrical leaf arrangement in the Marantaceae (M); f) homotropous; g) not encountered; h) antitropous. (f–h after Tomlinson 1961.)

The leaves on a plant often vary greatly in size and shape, some may be foliage leaves, some may be scale leaves (64), and this general phenomenon of variability is described as leaf polymorphism or heterophylly, although the latter term is perhaps better retained to apply specifically to changes in leaf form induced by the environment. If the plant has two very distinctive types of leaf the condition is described as dimorphism (30). In other cases two leaves of different size or shape occur at the same node, this arrangement being described as anisophylly (32). In addition all plants show at some stage in their development a changing progression of leaf shape, this sequence being described as a heteroblastic series such as almost inevitably occurs along the seedling axis of the plant (28, 29a), and often is also present along any developing lateral shoot (29d). For example the first leaves on axillary shoots might be scale leaves, each leaf being slightly more elaborate than the previous. This might give way gradually to a sequence of foliage leaves, and then the sequence may revert back to the production of scale leaves similar to those at the proximal end of the shoot. Such a shoot might then terminate in an inflorescence, each flower of the inflorescence being subtended by a bract (62) which is in itself a form of scale leaf.

Fig. 28. *Albizzia julibrissin*
A heteroblastic sequence consisting of a pair of simple cotyledons followed by a once-pinnate foliage leaf and two bipinnate foliage leaves. This seedling has the sequence out of step as the second bipinnate leaf is less elaborate than the first.

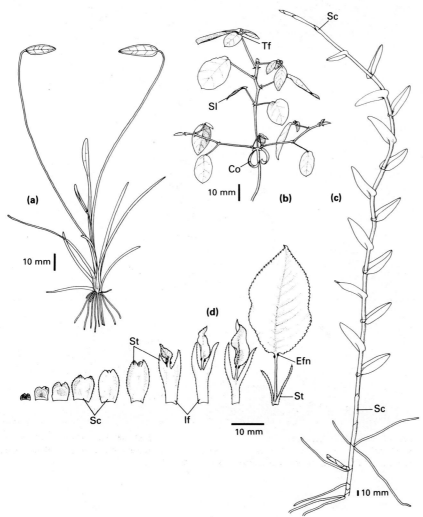

Fig. 29. a) *Alisma plantago*, seedling; b) *Kennedia rubicunda*, seedling; c) *Epidendrum ibaguense*, single shoot; d) *Prunus avium*, leaf sequence on developing shoot. Co: cotyledon. Efn: extra-floral nectary (80). If: intermediate form. Sc: scale leaf. Sl: simple leaf. St: stipule (52). Tl: trifoliate leaf.

One of the most obvious types of heterophylly (different leaf forms on the same plant 28) is that of dimorphism. This is the production of two totally different shapes of leaf during the life of the plant. The phenomenon is true of most plants in the sense that the cotyledons are usually distinct in form to subsequent leaves (cf. onion 163e) and likewise many plants bear scale leaves (64) on perhaps rhizome, bud, or in association with flowering (bracts 62). However, some plants illustrate an abrupt change of leaf form associated with environment such as occurs in water plants where there may be a submerged leaf form and an aerial leaf form. Similar abrupt changes in leaf form can occur in the aerial system of plants (31c) or between juvenile and adult portions of the plant (31a, b; 243).

Fig. 30. *Acacia pravissima*
A young seedling axis showing sudden transition from bipinnate leaves below, to simple phyllodes (42) above.

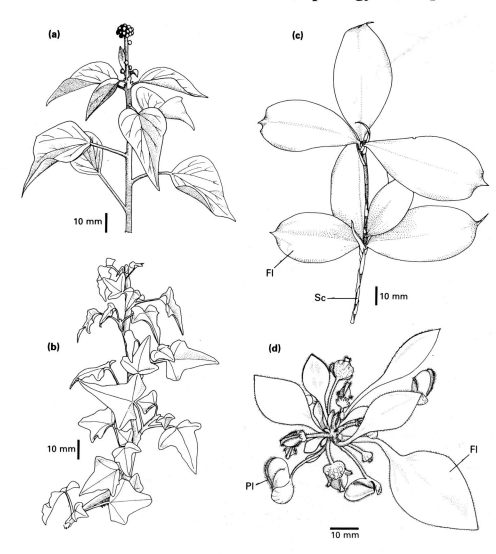

(a)

10 mm

(b)

10 mm

(c)

Fl

Sc

10 mm

(d)

Pl

Fl

10 mm

Fig. 31. a) *Hedera helix*, adult form; b) *Hedera helix*, juvenile form; c) *Dracaena surculosa*, aerial shoot; d) *Cephalotus follicularis*, seedling from above. Fl: foliage leaf. Pl: pitcher leaf (88). Sc: scale leaf (64).

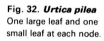

Fig. 32. *Urtica pilea*
One large leaf and one
small leaf at each node.

Anisophylly is a term commonly applied to a
condition of heterophylly (28) in which different
sized or shaped leaves occur at the same node
(i.e. nodes with opposite phyllotaxy **219i**).
However, its use is appropriate in any situation
where a difference of leaf form or size is repeated
on a regular basis. On horizontal shoots with
opposite and decussate phyllotaxy (**219j**) the
leaves of the lateral pairs are likely to be of the
same size whilst the leaves of the dorsiventral
pairs will be unequal in size. There is evidence
that such anisophylly can be primary, resulting
from an irreversible difference in leaf primordium
size from the start, or secondary, dependent upon
shoot orientation at the time of development of
the leaf pair. Anisophylly can also occur in
horizontal (plagiotropic 246) shoots in which
there is only one leaf per node. In this situation,
leaves borne on the upper side of the shoot will
be of a different size (usually smaller) than those
on the lower side. This type of anisophylly is
sometimes referred to as lateral anisophylly
distinguishing it from the nodal anisophylly
recorded above. Plants in which all leaf pairs are
anisophyllous may show an overall symmetry of
pattern within a branch complex which is
particularly apparent if the leaves themselves are
asymmetrical (**33e**). Frequently the activity or
potential (242) of the bud or buds in the axil of
the larger leaf of an unequal pair is greater or
different compared with the bud in the axil of the
smaller leaf (anisoclady **33a, b, c**). Anisophylly
can occur even at the cotyledon stage
(anisocotyly **163f, 209**).

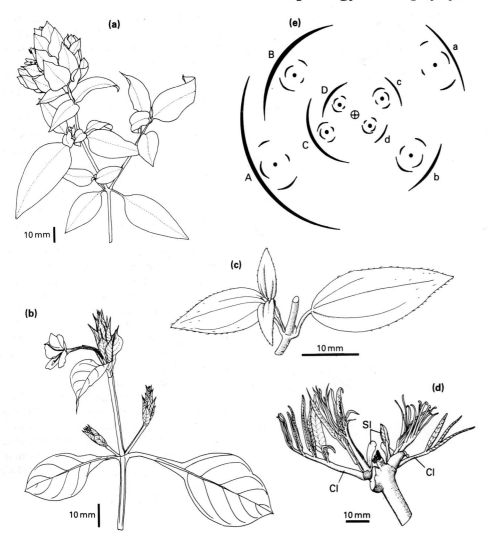

Fig. 33. a) *Beloperone guttata*; b) *Eranthemum pulchellium*; c) *Monochaetum calcaratum*, single node; d) *Phellodendron lavallii*, developing shoot pair; e) plan view of shoot system of *B. guttata* (a) indicating symmetry of large and small leaf at each node, e.g. Aa, Bb. Cl: compound leaf. Sl: simple leaf.

10 mm

10 mm

10 mm

10 mm

The veins (i.e. anatomically speaking the vascular bundles) form prominent features of many leaves. The pattern of venation is often distinctive for a given plant species or may be characteristic for a larger taxonomic group. A classification for the venation of leaves, and also their shapes, is given in detail in Hickey (1973). Generally monocotyledons are described as having parallel venation reflecting the insertion of the leaf base all round the stem, and to lack free vein endings. Parallel veins are inevitably joined by numerous fine cross-connections (**22**). However, there are numerous exceptions (**34, 35e**). In contrast dicotyledons are described as having reticulate (network) venation (**35a**) and a considerably range of basic patterns is to be found. Nevertheless a number of dicotyledons have a parallel venation (**35c, d**) or very rarely dichotomous venation. The veins of an individual leaf can usually be categorized into primary veins, secondary veins, and so on; and a marginal vein may be prominent. A number of examples are illustrated here (**35g**). The areas bounded by the ultimate veins are termed areoles (**35f**) (cf. Cactaceae 202) and the blind ending veinlets entering these again make distinctive patterns.

Fig. 34. *Dioscorea zanzibarensis*
A monocotyledon with net-veined leaves.

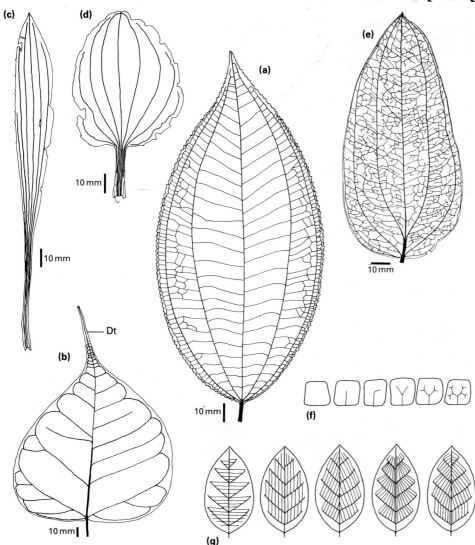

(c)

(d)

10 mm

10 mm

10 mm

(a)

(b)

— Dt

10 mm

10 mm

(e)

10 mm

(f)

(g)

Fig. 35. a) *Clidemia hirta* (D); b) *Ficus religiosa* (D); c) *Plantago lanceolata*, parallel veined (D); d) *Plantago major*; e) *Smilax* sp., net-veined (M); f) typical areole patterns showing ultimate veinlets; g) variations of one type of secondary vein layout. Dt: drip tip. (f and g after Hickey 1973). (D)=dicotyledon, (M)=monocotyledon.

Developing leaves are often confined within a protective structure, a bud (264), and at this stage will quite possibly have acquired more or less their final shape (mostly by cell multiplication) but not reached their final size (mostly due to cell enlargement). Depending upon the number, size, and complexity of the leaves in the bud, they are likely to be variously folded, the manner of folding being consistent for any given species. The phenomenon of leaf packing is referred to by a variety of terms. The contortion of a single leaf is called ptyxis (from the Greek for folding), the various modes of packing of leaves together is referred to as vernation (or prefoliation 38). The packing of perianth segments in a flower bud is very similar to that of vegetative leaves in a vegetative bud and is termed aestivation or prefloration (148). The different forms of ptyxis are frequently of diagnostic value in identifying a plant and as such have acquired an extensive range of terminology. The most common terms are illustrated here (37) in three-dimensional diagrams which convey more information than an over-simplified written description. The individual leaflets of a compound leaf may be folded in one manner, the leaf as a whole showing an alternative arrangement. The folding of the leaves of palms is particularly elaborate and is a function of their unique mode of development (92). An extensive discussion of ptyxis occurs in Cullen (1978).

Fig. 36a. *Drosophyllum lusitanicum*
An insectivorous plant with circinate ptyxis. Unusual in that the leaves are rolled outwards rather than inwards as in Fig. **37e.**

Fig. 36b. *Nelumbo nucifera*
Peltate leaves (88) with involute folding (**37b**).

Fig. 37. Types of individual leaf folding. a) curved; b) involute; c) revolute; d) supervolute (also termed convolute, cf. **39e, f**); e) circinate (cf. **36a**); f) supervolute/involute; g) conduplicate/involute; h) conduplicate/plicate; i) plicate; j) conduplicate; k) explicative; l) plane (flat).

Fig. 38. *Rhizophora mangle*
Convolute vernation (**39f**) of a pair of opposite leaves.

Individual leaves are variously folded (ptyxis 36) and variously packed in a bud (vernation or prefoliation). The manner of packing is often distinctive and is a noticeable feature of the perianth segments in the case of flower buds (aestivation 148) to which the terms illustrated here also apply. The form of vernation depends to some extent on the number of leaves at a node (phyllotaxis 218). In monocotyledons where there is only ever one leaf at a node, leaves are likely to be folded or rolled, if at all, in a manner consistent with the protection of each leaf by the sheath of the preceding or axillant leaf or by a more or less tubular prophyll (66a). In more condensed shoot apices, where the phyllotaxis is likely to be spiral, overlapping of adjacent leaves will be more elaborate. In dicotyledons, particularly those with two or more leaves per node, a variety of formats are found. The edges of adjacent leaves may not touch (open vernation **39c**) or just touch but not overlap (valvate **39b**). Two leaves at a node may face each other and then be appressed (**39a**) or opposite (**39d**). Overlapping leaves (or petals) are said to be imbricate (e.g. Fig. **149d–j** for perianth segments). Care must be taken to identify the precise details of imbrication; sectioning of unopened buds, and close scrutiny of a series of buds as they unfold, may be necessary. A common form of imbrication in flowers, but rare in vegetative buds, is convolute (**39e**) (cf. the use of this term in ptyxis **37d**). Convolution may occur even where only two leaves are involved (**38, 39f**). When the individual leaves are conduplicate (**37j**) then the vernation may be equitant (**39g**) or obvolute (half-equitant **39h**). As a bud expands, distinctive patterns of colour or ridging may be seen on one leaf due to the pressure exerted by a neighbouring enlarging leaf. Such markings are quite common in monocotyledons with linear leaves, such as grasses, and are termed constriction bands.

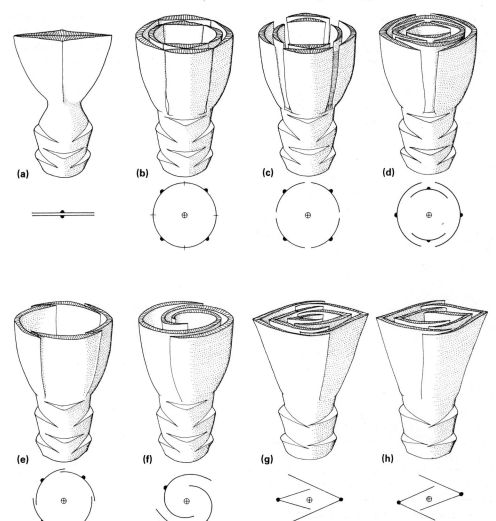

Fig. 39. Folding of leaves together. a) appressed, b) valvate, c) open, d) opposite, e) convolute, f) convolute, g) equitant, h) obvolute.

The leaves of many dicotyledons and some monocotyledons (e.g. Araceae) have a stalk or petiole separating the leaf blade from its base or point of attachment to the stem. The development of the petiole is different in these two groups (20). Leaves lacking petioles are said to be sessile. Occasionally the leaf blade is apparently absent and the petiole flattened laterally into a photosynthetic organ—a phyllode (42). Likewise in a leaf bearing a lamina, the petiole may bear wings along its side (49d). The petioles of a number of climbing plants are persistent, remaining on the plant and becoming woody after the lamina has become detached (41h), the abscission zone developing into a point. More frequently the petiole will fall with the leaf with or without an abscission joint (48). Petioles may be fleshy and swollen (41f) or sensitive to touch acting as twining supports in climbing plants (41e, h). The whole or part of a petiole may form into a permanent woody spine (40b, 41a, b, c, d). The orientation of a leaf lamina can be affected by movement in the petiole. This can take the form of twisting of the petiole (238) or may be due to the presence of one or more pulvini (46) or due to the presence of a pulvinoid (46). This latter structure is a swelling similar to a pulvinus but operates only once producing an irreversible repositioning of the lamina. The proximal end of a petiole frequently forms a protective cavity surrounding the axillary bud (265b) or the terminal bud (265b, d). The cavity formed by longitudinal folding of petioles of *Piper cenocladium* is inhabited by ants; food cells develop on the inner surface of

this cavity. Similar food bodies occur on the petiole of *Cecropia* (78). In *Vitellaria* the petioles of the cotyledons elongate at germination forcing the radicle of the seedling underground (41g).

Fig. 40a. *Psammisia ulbriehiana*
Long-lived leaf with woody petiole being encroached upon by stem expansion.

Fig. 40b. *Quisqualis indicus*
Petiole spines (laminas shed).

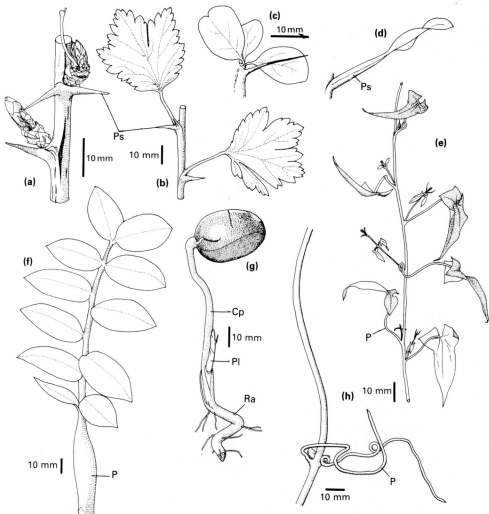

Fig. 41. a, b) *Ribes uva-crispa*; c, d) *Fouquieria diguetii*, spine formed from stem tissue adnate (234) to abaxial surface of petiole (d); e) *Maurandia* sp.; f) *Zamioculcas zamifolia*, single leaf; g) *Vitellaria paradoxum*, germinating seedling; h) *Clematis montana*, single node. Cp: cotyledonary petiole. P: petiole. Pl: plumule. Ps: petiole spine. Ra: radicle.

Many leaves, especially of dicotyledonous but also of certain monocotyledonous plants, may be described in terms of sheath (50), petiole (40), and lamina (22), although the development of these structures in the two groups is different (20). A 'typical' petiole of a dicotyledonous leaf is a cylindrical, not necessarily photosynthetic, stalk. In a number of plants, however, the petiole is flattened and contributes considerably to the light interception area of the leaf. The lamina may appear to be correspondingly rudimentary (45c) or apparently absent as inferred by a transition series (45). Such apparently flattened petioles are termed phyllodes, and the flattening may be dorsiventral (53c) or more often lateral (43). Recent developmental studies suggest that the phyllode can represent a modification of the whole leaf, not just the petiole (44). A phyllode, being a leaf, will subtend a bud or shoot and may thus be distinguished from flattened stems, cladodes and phylloclades (126), which are themselves subtended by leaves.

Fig. 42. *Acacia paradoxa*
Each phyllode subtends a number of buds (236) one of which develops into an inflorescence. Stipular spines (56) present (cf. Fig. **43a, c**).

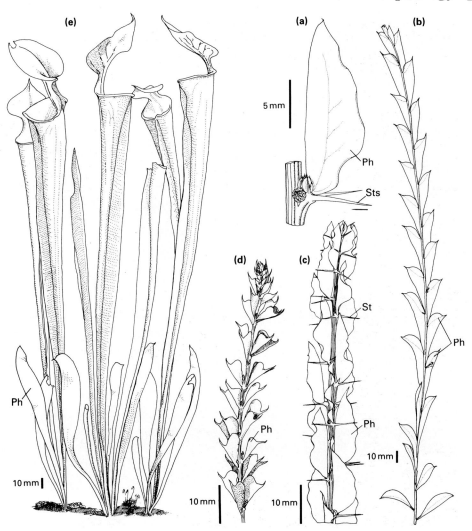

Fig. 43. a) *Acacia paradoxa*, single node; b) *A. glaucoptera*, leaves on seedling axis; c) *A. paradoxa*; d) *A. pravissima*; e) *Sarracenia flava*, phyllodes and ascidiate (88) leaves. Ph: phyllode. Sts: stipular spine (56).

In order to compare the nature of apparently similar structures in different plant species it is usually helpful, if not essential, to compare their development. This is particularly true in the case of leaf structure (18) and has aided the comparison of dicotyledonous leaves and monocotyledonous leaves (20). It has also proved useful in the interpretation of phyllode structure (42). Thus developmental studies of the phyllode of species of *Oxalis* indicate that the lamina is suppressed, more so in some leaves than others, and that the phyllode does indeed represent a flattened petiole. This is not true however for those *Acacia* species (**43a, b, c, d**) bearing phyllodes (Kaplan 1975). In these cases the whole rachis (22) of the leaf is involved in the formation of the phyllode, developing as a flattened structure due to the activity of an adaxial meristem (18). All the leaves on an *Acacia* plant (except the seedling leaves 30) may develop in the manner of phyllodes; in some species a variable range of 'transitionary' forms will be found. Thus the older (proximal) branches of *Acacia rubida* bear bipinnate leaves (**45k**) whilst the younger (distal) branch ends will bear phyllodes (**45a, b**). In between a range of intermediate types may be found (**45c–j**). Adaxial meristem activity can occur anywhere along the petiole/rachis axis and is combined with a variable reduction in the activity of the primordia responsible for leaflet production. The compound leaf of *A. rubida* is paripinnate (no terminal leaflet 22) but this leaf, and the phyllodes, and the intermediate forms, bear a minute terminal 'pointlet', representing the distal oldest end of the leaf (**45j**). For the *Acacia spp.* therefore, at least, the phyllode does not represent a flattened petiole, but rather a flattened rachis as determined by developmental studies.

Fig. 44. *Acacia rubida*
Phyllodes and bipinnate leaves. A 'transitional' leaf at centre.

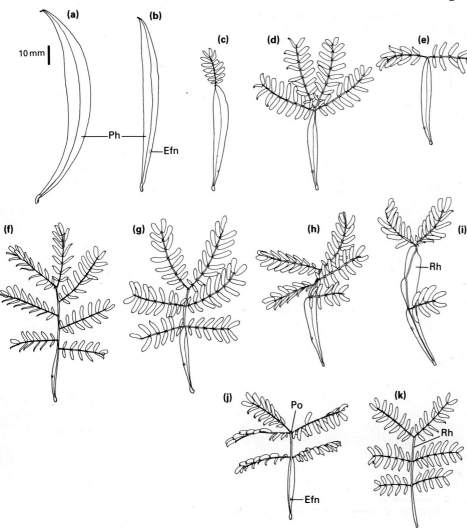

Fig. 45. A selection of leaf forms from one tree of *Acacia rubida*. a, b) are adult foliage types; k) is the juvenile form; c)–j) are intermediate types. Efn: extra-floral nectary. Ph: phyllode. Po: pointlet. Rh: rachis.

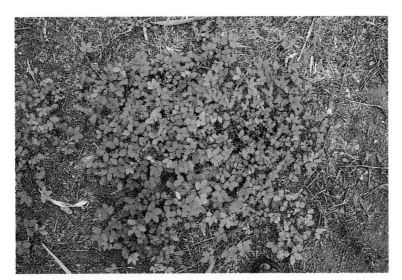

Fig. 46a, b. *Mimosa pudica*
a) Undisturbed plant; b) 5 seconds after disturbance. Pulvini at base of leaves and leaflets have distorted, acting as hinges causing the leaves to collapse. Most pulvini operate much more slowly.

Localized swellings of leaf or leaflet stalk are common occurrences in both dicotyledons (**47**) and monocotyledons (**220**). Usually such swellings act as hinges, allowing more or less reversible movement between the parts of the leaf (**46a, b, 47a, a′**). Such hinges are referred to as pulvini (singular pulvinus) and may be found at the base of the petiole (**80b**), or at the junction of petiole with lamina (**220**), and/or at the base of each leaflet in a compound leaf (**47d**). They also occur on stems (128). Displacement of the leaf with respect to light or gravity will cause a compensatory reorientation of the leaf parts as the cells at one side of the circumference of the pulvinus swell or shrink due to water gain or loss. Abscission joints (48) and pulvinoids often closely resemble pulvini. Pulvinoids form irreversible growth joints reorientating a leaf or leaflet once only or forming a clasping aid to climbing (**41h**). Abscission joints locate the point of weakness where a leaf or leaflet or portion of petiole or rachis will eventually break, and are not capable of reorientation but are usually identified by an associated annular groove (**49a**). As viewed in a transverse section, the anatomy of a pulvinus will differ from that of a pulvinoid or an abscission joint: pulvinus—reversible movement possible, vascular bundles located centrally and often lignified; pulvinoid—irreversible movement, no groove present, vascular bundles peripheral and not lignified; abscission joint—no movement possible, groove present. The swelling at each node on the culm (180) of a grass plant is in fact a pulvinus at the base of the leaf sheath inserted at that node.

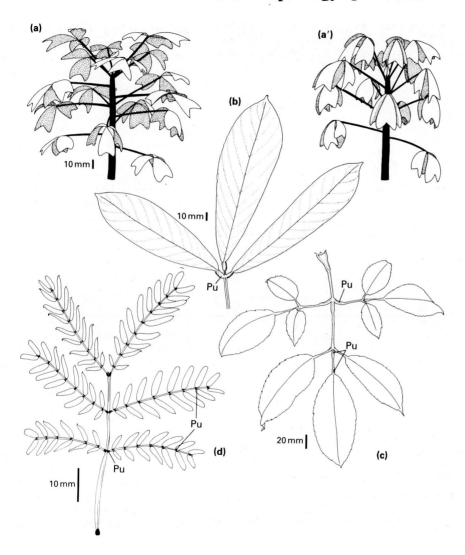

Fig. 47. a, a¹) *Oxalis ortgeisii*, day and night leaf arrangement; b) *Derris elliptica*, end of compound leaf; c) *Leea guineense*, single leaf; d) *Acacia heterophylla*, single leaf. Pu: pulvinus.

The construction of a leaf is often articulated, i.e. jointed, and the leaf will eventually fall apart at the points of articulation—the abscission joints (or struma). Such joints are frequently swollen (**49a, c**) and usually also bear an annular constriction groove marking the location of future breakage (**49a**). Abscission joints may occur at intervals along the rachis of a compound leaf and/or at the base of each individual leaflet, or simply near the base of the leaf itself. Abscission joints resemble pulvini (46) and pulvinoids (46) but do not reorientate the leaf or leaflet. Pulvini may be present in addition to abscission joints and then the pulvinus may be shed with the leaf or left behind (**48**). The point of breakage resulting in leaf fall (i.e. the abscission zone) is not necessarily recognizable externally as a swollen abscission joint. Indeed in some trees the leaf is only forced to fall by the increase in girth of the stem bearing it (**40a**). Similar zones of breakage occur in stems (see cladoptosis 268).

Fig. 48. *Philodendron digitatum*
A palmate leaf, two leaflets detached, breakage occurring at the mid-point of the pulvini.

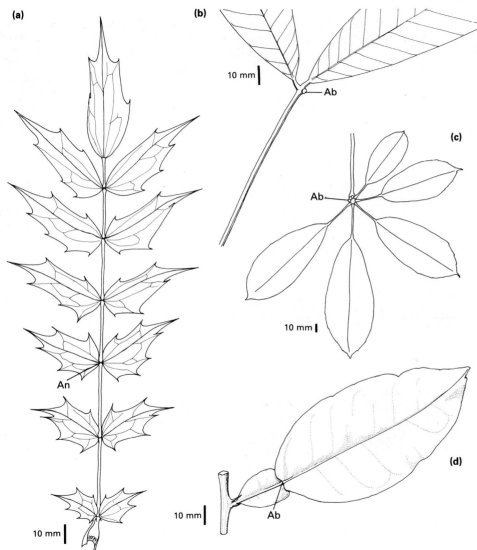

(a)

(b)

10 mm

Ab

(c)

Ab

10 mm

An

10 mm

10 mm

(d)

Ab

Fig. 49. a) *Mahonia japonica*, single leaf; b) *Hevea brasiliensis*, part of leaf; c) *Schefflera actinophylla*, single leaf; d) *Citrus paradisi*, single node. Ab: abscission joint. An: annular groove.

Fig. 50. *Hedychium gardnerianum*
The leaf sheath to the left of the fruit clasps the stem; the one below is pulled away from the stem.

The structure of a leaf often lends itself to description in terms of leaf blade (or lamina 22), leaf stalk (or petiole 40), and leaf base (or sheath). These terms are applied equally to dicotyledonous and monocotyledonous leaves, although the development of leaves in these groups is fundamentally different and the petiole of one is only equivalent to the petiole of the other in a purely descriptive sense (20). Sheath is perhaps the least well defined of these descriptive terms and has been applied to any structurally distinctive portion of the leaf at or near the point of insertion of the leaf on the axis. Sheaths sometimes bear stipules (52). Leaf sheaths range in structure from barely noticeable enlargements of the base of the petiole (**33c**), to prominent elaborations clasping the stem (**50, 51b, c**). A sheath may partially or totally protect the axillary bud (**51c, d**). Either the proximal or the distal end of the sheath can be modified into a pulvinus (46), or may become persistently woody or fibrous (**51b**) as in many palms. A leaf sheath is a particularly conspicuous feature of most monocotyledonous leaves, encircling the stem due to the mode of development of these leaves (18). An aggregate of such concentric leaf bases forms a pseudostem of, for example, a banana (*Musa* sp.). In some instances the leaf base forms the bulk of the photosynthetic surface of the leaf (**89c**).

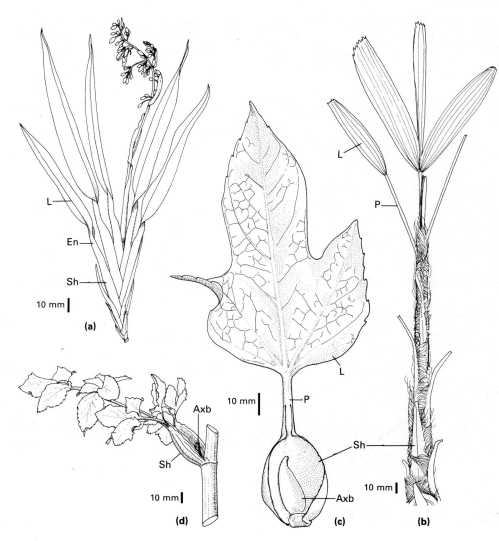

Fig. 51. a) *Dianella caerulea*, top of aerial shoot; b) *Rhapis excelsa*, top of aerial shoot; c) *Fatsia japonica*, single leaf; d) *Smyrnium olusatrum*, single node. Axb: axillary bud. En: ensiform portion of leaf (86). L: lamina. P: petiole. Sh: sheath.

Fig. 52. *Liriodendron tulipifera*
The pair of stipules at the base of each leaf petiole protects the next youngest leaf here seen silhouetted inside.

A stipule is an outgrowth associated with the base of a leaf developing from part of the leaf primordium in the early stages. Plant species are termed stipulate or exstipulate (with or without stipules). Stipules are not common in monocotyledons where they usually occur one per leaf (**55b**) or very occasionally two per leaf (**57b**). Stipules of dicotyledons are paired typically one on either side of the point of insertion of the petiole on to the stem (**53b**). However, there are many positional variations often including fusion of structures (**54**). Stipules may be relatively small and insignificant (**53c, d**), often scale-like (**61b, 80b**), and may fall off early in the life of the leaf leaving a scar (**78**). They often protect younger organs in the bud (**52**), and then fall when the bud develops. Conversely stipules can be very conspicuous and leaf-like (**55a, 57e**) or resemble entire leaves (**55d, 69e**) from which they may be recognized by the absence of associated axillary buds. It is quite possible that in some cases structures traditionally described as stipules (as in many members of the Rubiaceae (**55d**), see for example Rutishauser 1984) in fact represent whole leaves. The structures in question themselves bear outgrowths (colleters 80) which could be rudimentary stipules and thus all the members of the whorl of 'leaves' at a node would be foliar in origin, some with axillary buds, some without. Stipules may be modified into a number of structures (**56**), especially spines (**43c, 6**). These are lignified (woody) and usually persist after the rest of the leaf has fallen.

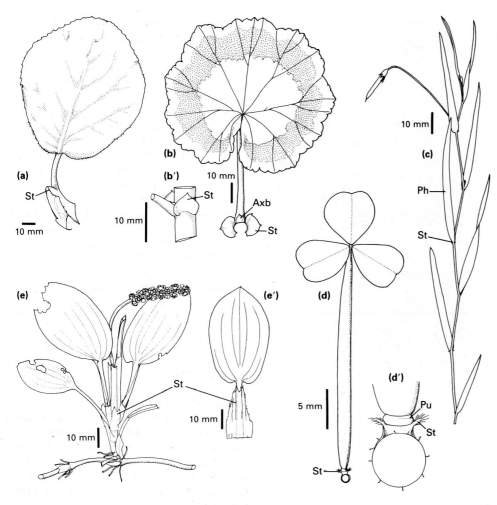

Fig. 53. a) *Bergenia* sp., single leaf; b, b¹) *Pelargonium* cv., single leaf and side view node; c) *Lathyrus nissola*; d, d¹) *Oxalis* sp., single leaf and top view node; e, e¹) *Potamogeton* sp., leaf rosette and single leaf. Axb: axillary bud. Ph: phyllode (42). Pu: pulvinus (46). St: stipule.

(a)

St

10 mm

(b)

(b′)

St

St

10 mm

10 mm

Axb

St

(c)

10 mm

Ph

St

(d)

(e)

(e′)

St

10 mm

10 mm

10 mm

5 mm

St

(d′)

Pu

St

St

1 mm

Fig. 54. *Reynoutria sachalinensis*
The stipule, an ochrea, completely encircles the stem.

Adventitious root primordia (98) visible just below the node.

The position of a pair of stipules in dicotyledonous plants relative to the leaf base, petiole, and insertion of the leaf on to the stem, varies considerably. The stipule may be located at the extreme proximal end of the petiole (**53b, 55f**) or be borne actually on the stem apparently detached from the leaf base (**55g**). The stipules can also be found at the junction of the leaf base (sheath) and petiole (**55h, i**). The stipule is described as adnate if it is fused along part of the length of the petiole (**55f, k**). A stipule may be attached to the side of the stem, 90° around from the point of leaf insertion (interpetiolar **55i**) and may then be fused with the corresponding stipule of a second leaf at that node (**55c**). Stipules are occasionally found on the side of the stem away from the point of leaf insertion (**55m**), a situation described as 'abaxial' or 'counter', or more precisely as leaf opposed. Conversely a single stipule may be found in a truly adaxial position between the petiole and the stem (**55a, j**): a median or intrapetiolar stipule. A similarly located single structure often described as a stipule is found in a number of monocotyledonous plants (**55b, 53e**), that of *Eichhornia* being particularly elaborate bearing an additional terminal structure, the stipular lobe. If the single stipule encircles the whole stem it is described as an oc(h)rea (**55e, n, 54**).

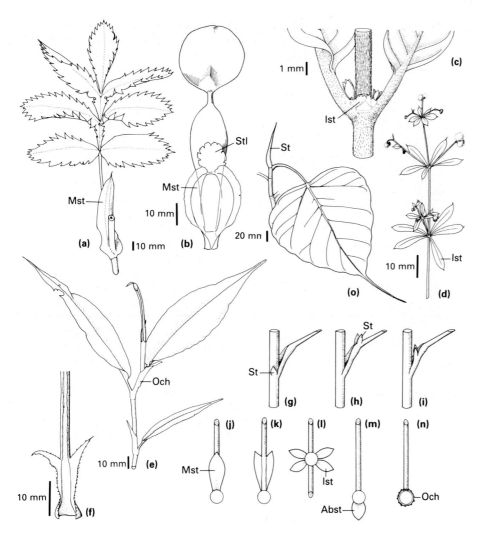

Fig. 55. a) *Melianthus major*, single leaf at node; b) *Eichhornia crassipes*, single leaf; c) *Manettia inflata*, leaf pair at node; d) *Galium aparine*; e) *Polygonum* sp.; f) *Rosa* sp., base of leaf petiole; o) *Ficus religiosa*, young end of shoot. g)–n) range of stipule locations. Abst: abaxial stipule. Ist: interpetiolar stipule. Mst: median (intrapetiolar) stipule. Och: ochrea. St: stipule. Stl: stipular lobe.

Members of the genus *Smilax* (Liliaceae) are most unusual amongst monocotyledons in that each leaf bears two structures in a stipular position which are modified into tendrils (**57b**). Stipular structures in other monocotyledons occur singly and are usually membraneous or otherwise rather insignificant (**53e**). In dicotyledonous plants the stipules may persist as long as the rest of the leaf or may fall a long time before or after the leaf or not at all. Such persistent stipules are usually modified into the form of woody spines lasting for many years (**202b, 119f**). The stipular spines of some species of *Acacia* are hollow and inhabited by ants (**205a, 6**). One of each pair of stipular spines in *Paliurus spina-christi* is straight and the other is reflexed. Stipules also occur that are modified as extra-floral nectaries (**56**), or are represented by a fringe of hairs (**57c**). In many plants the scale-like stipules of leaves aggregate in a dormant bud performing a protective role (**265c**).

Fig. 56. *Bauhinia* sp.
Stipules represented by extra-floral nectaries (80).

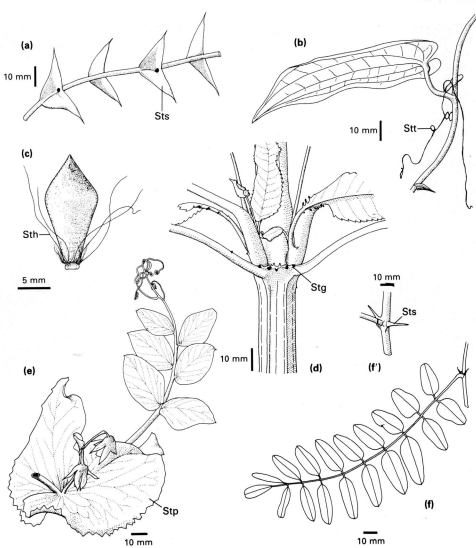

(a)

10 mm

Sts

(b)

10 mm

Stt

(c)

Sth

5 mm

(e)

Stp

10 mm

(d)

Stg

10 mm

(f')

10 mm

Sts

(f)

10 mm

Fig. 57. a) *Acacia hindisii*, woody stipular pairs; b) *Smilax lancaefolia*, single node; c) *Anacampseros* sp., single leaf; d) *Impatiens balsamina*, leaf pair at node; e) *Pisum sativum*, single leaf at node; f, f¹) *Robinia pseudacacia*, single leaf at node and close view node (compare **119f**). Sth: stipular hairs. Stg: stipular glands. Stp: photosynthetic stipule. Sts: stipular spine. Stt: stipular tendril.

Fig. 58. *Phaseolus coccineus*
A pair of small stipels located on the leaf petiole just below its junction with the lamina.

Occasionally the individual leaflets in a compound leaf have small outgrowths at their bases resembling stipules (52). These structures are referred to as stipels (or secondary stipules, or stipella) and are usually uniform in size on the leaf (59c) or may vary considerably. They are most frequently met with in members of the Leguminosae (58, 59). Many compound leaves and some simple leaves (25c) have irregularly placed small leaflets, interspersed between major leaflets (interruptedly pinnate) which resemble stipels in appearance but lack the precise location at the proximal end of each main leaflet. A number of other structures can be found on a compound leaf in a similar position to those of stipels but not of a membraneous or leaf-like appearance. Examples include prickles (77d, e) and extra-floral nectaries (81d).

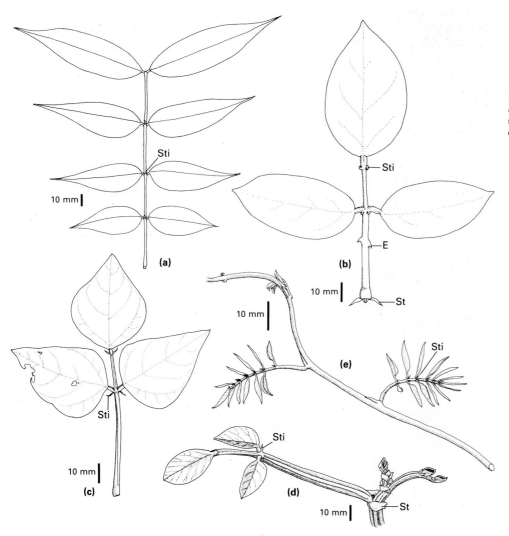

Fig. 59. a) *Cassia floribunda*, single leaf (cf. **81c**); b) *Erythrina crista-galli*, single leaf; c) *Phaseolus vulgaris*, single leaf; d) *Butea buteiformis*; e) *Wistaria sinensis*. E: emergence (76). St: stipule. Sti: stipel.

Sti

10 mm

(a)

Sti

E

10 mm

St

(b)

Sti

10 mm

(c)

10 mm

Sti

(e)

Sti

(d)

10 mm

St

Fig. 60. *Mutisia acuminata*
The proximal, i.e. lowest, pair of leaflets of the pinnate leaf is located in a stipular position. Family Compositae—mostly exstipulate.

In the compound leaves of some dicotyledons, the proximal pair of leaflets is positioned very close to the point of insertion of the leaf at the node and thus appears to be in the stipular position. True stipules may also be present (**61b**) in which case the nature of the basal leaflets is apparent. If not, these leaflets are sometimes referred to as pseudostipules. These leaflets are also termed pseudostipules if the family to which the plant belongs is predominantly exstipulate (**60**). The pseudostipules may have a different shape to other leaflets on the same leaf (**61a**). A careful study of the development of the leaf primordia may indicate the relationship of the lowest pairs of leaflets to those located more distally, revealing that they are indeed pseudostipules rather than stipules. In some cases the nature of the vascular supply to the leaf and pseudostipules may be of value. Stipulate leaves often have three leaf traces, exstipulate leaves often only one. The single prophyll of some *Aristolochia* spp. (**67c**) is sometimes referred to as a pseudostipule.

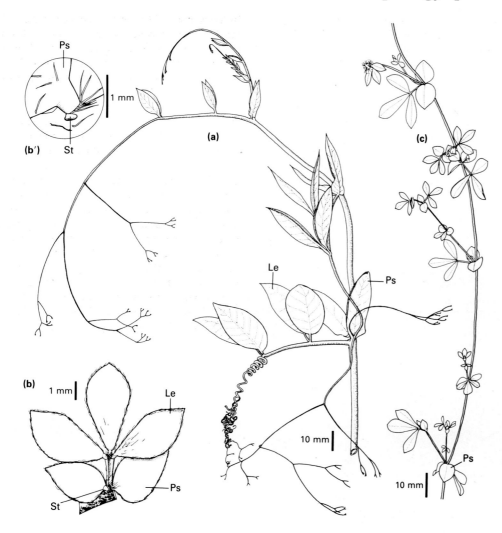

Fig. 61. a) *Cobaea scandens*, end of climbing shoot; b, b¹, c) *Lotus corniculatus*, c) portion of shoot, b) single leaf, b¹) close view of minute stipule. Le: leaflet. Lt: leaf tendril. Ps: pseudostipule. St: stipule.

Leaves located in association with flowers are frequently modified or reduced in size relative to vegetative leaves on the same plant. Such leaves are referred to as bracts (or hypsophylls cf. cataphylls 64). Any leaf, modified or not, that subtends a flower can be termed a bract although there are many instances in which flower buds are found without associated subtending leaves. Conversely the stalk (pedicel) of an individual flower may bear a bract (typically one in monocotyledons and two in dicotyledons, 66) which may or may not subtend its own flower. Such a leaf is termed a bracteole. Thus the bracteole of one flower may be the bract of another flower (63e). If a number of flowers are borne in a condensed inflorescence, their individual bracts will occur in a tight whorl or involucre (144). However, an involucre may be associated with a single flower (147e). A compound umbel (141m) will display an involucre of bracts at its base and an involucel beneath each distal flower cluster. Each individual bract of an involucre can be called a phyllary. One or several of the bracts associated with an inflorescence may be relatively large and conspicuous (62a, 63a, d). Such bracts may assist in the wind dispersal of fruit or fruits (235e). Bracts are a conspicuous feature of many monocotyledonous inflorescences (63a, c) and form distinctive features of the grass spikelet (186). Generally, bracts may appear leaf-like, are frequently scale-like, may be massive as in many palms, or modified into spines (62b), hooks (161b), or persistent woody structures surrounding fruits (157h, 155o; the fused woody bracts—cupule—as in oak for example).

Fig. 62a. *Cephaelis poepiggiana*
A pair of coloured bracts beneath an inflorescence.

Fig. 62b. *Barleria prionitis*
Young flower buds, in axils of foliage leaves, surrounded by bracts in the form of spines (i.e. leaf spines 70).

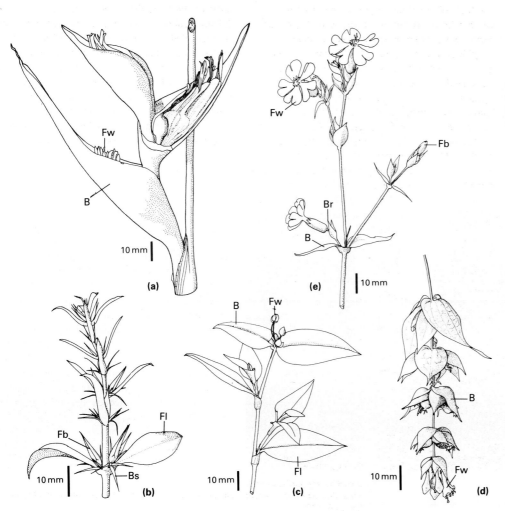

Fig. 63. Portions of inflorescences incorporating bracts. a) *Heliconia peruviana*, b) *Barleria prionitis*, c) *Tradescantia* sp., d) *Leycesteria formosa*, e) *Silene dioica*. B: bract. Br: bracteole. Bs: bract spine. Fb: flower bud. Fl: foliage leaf. Fw: flower.

Fig. 64. *Agave americana*
Scale leaves on the extending flowering axis (see Fig. opposite the Introduction for mature inflorescence).

A great many plants are dimorphic (30), bearing membraneous scale leaves in addition to relatively large foliage leaves intercepting light. These 'cataphylls' are sometimes devoid of chlorophyll, and often perform a protective role surrounding vegetative or floral meristems (**64, 62b**). Underground stems of rhizomatous plants commonly bear scale leaves (**65a, e,** but cf. **87c**) which may or may not subtend axillary buds. Successive leaves located along a shoot may demonstrate a heteroblastic series (**29c**) from a simple scale leaf to a more or less elaborate foliage leaf. A similar heteroblastic sequence occurs in relation to flowering shoots, the foliage leaves at the proximal end of the inflorescence merging into scale leaves at the distal end. Scale leaves associated with an infloresence are termed hypsophylls or more commonly bracts and bracteoles (**62**). Particularly in monocotyledons, the first leaf on a shoot (the prophyll **66**) is often represented by a cataphyll and differs greatly in size and morphology to more distal leaves on that axis. Scale leaves are typically smaller in size than the corresponding foliage leaves of a particular plant, although small is a relative term, the protective scale leaves (bracts) of the inflorescence of some palms being massive woody structures over 1 m in length.

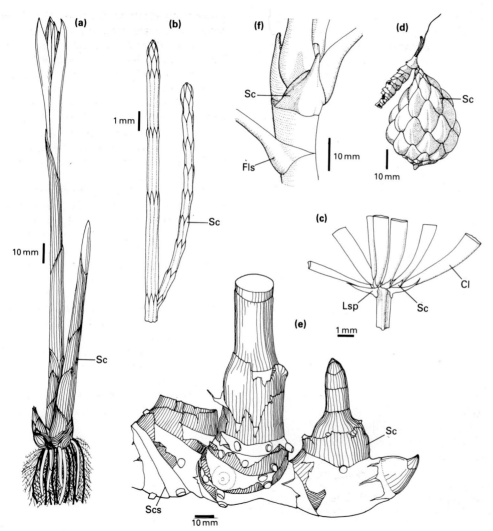

(a)

(b)

(f)

(d)

1 mm

10 mm

Sc

10 mm

Fls

Sc

10 mm

(c)

Cl

Lsp   Sc

Sc

(e)

1 mm

Sc

10 mm

Scs

10 mm

Fig. 65. a) *Cyperus alternifolius*, developing aerial shoots; b) *Casuarina equisetifolia*, distal end of shoot; (c) *Asparagus densiflorus*, single node (cf. **127a**); d) *Raphia* sp., fruit; e) *Costus spiralis*, rhizome; f) *Fatsia japonica*, scale leaves beneath shoot apex. Cl: cladode (126). Fls: foliage leaf sheath (**51c**). Lsp: leaf spine (70). Sc: scale leaf. Scs: scale leaf scar.

The term prophyll is applied to the leaf or leaves at the first (proximal) node on a shoot. The leaves in this position are often but by no means always represented by cataphylls (64) whether or not subsequent leaves are similarly modified. The single prophyll of many monocotyledons can be a particularly distinctive scale (66a), often appearing double with a double tip (bicarinate). It is almost always found in an adaxial (4) (or adossiete) position, i.e. on the top of the lateral shoot. Single adaxial prophylls also occur in some dicotyledons (67c, d). In dicotyledons prophylls of a pair are usually positioned laterally (66b, 67a, b); if only one is present it is not necessarily adaxial. A bracteole (62), being the first leaf on a shoot, is thus also a prophyll. The palea of a grass spikelet (186) and the utricle of a sedge (196) are likewise prophylls because of their positions. The prophyll on the shoot system forming a female inflorescence of *Zea* (190) occurs as the first of a series of large protective 'husks'. Prophylls are occasionally persistent and woody, represented by hooks, spines (203b, 71c), or possibly modified into tendrils (123e). The prophyll(s) may be involved in bud protection (264). In the case of sylleptic growth (262) the prophyll is separated from the parent shoot by a long hypopodium (263a). However, if the prophyll is inserted on the side shoot in a proximal position very close to the parent axis, then the bud in the axil of the prophyll can also develop very close to the parent axis and this process may be repeated giving rise to proliferation (238).

Fig. 66b. *Simmondsia chinensis*
Each axillary shoot bears two small lateral prophylls at its base (one of each pair visible from this viewpoint) which protect the axillary buds (264).

Fig. 66a. *Philodendron pedatum*
Same plant as Fig. **10**, later development stage. The pale coloured prophyll is about to fall, having protected its axillary shoot (an elongating hypopodium 262).

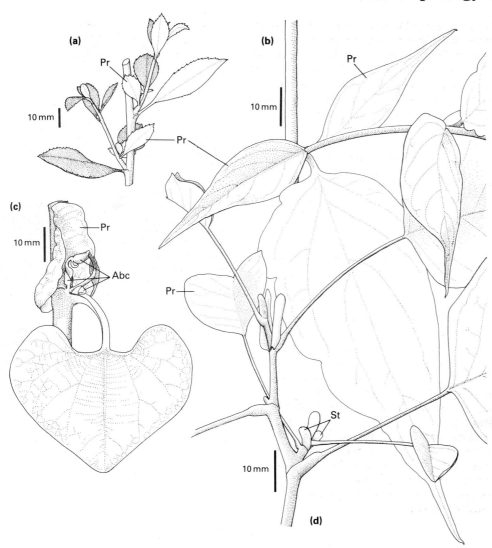

Fig. 67. a) *Escallonia* sp., b) *Leycesteria formosa*, prophyll pair at base of side shoot; c) *Aristolochia cymbifera*, single adaxial prophyll; d) *Liriodendron tulipifera*, single large prophyll. Abc: accessory bud complex (236). Pr: prophyll. St: stipule.

Fig. 68a. *Bignonia* sp.
One of the three leaflets of each leaf forms a persistent clasping woody tendril.

Climbing plants exhibit a considerable range of morphological features that prevent the shoot system falling. The stem may twine, may develop adventitious clasping roots (**98**), or tendrils and hooks which represent modified shoots (**122**) or inflorescences (**145b**), or may possess leaves all or part of which develop in the form of tendrils or hooks. In one genus only (*Smilax*) the stipules of the leaf operate in this manner (**57b**). The leaf petiole may be a twining organ (**41e, h**). Leaf tendrils themselves are found in a variety of forms. The distal extremity only of a simple leaf may be elongated forming a twining tendril (**68b, 69g**), or the whole leaf may be involved (**69e**). Alternatively either the terminal or one or more lateral leaflets of a compound leaf will occur as a tendril (**69a, b, c, f**). In the case of compound leaves, the proportion of leaflet tendril to ordinary leaflets may be flexible in a given species or the leaves may be very consistent in this respect. Tendrils show pronounced movement and will twine around a support once contact is made, usually due to faster growth rates on the side away from the support. In some species the encircling portion of leaf will subsequently become enlarged and woody and permanent (**68a**). A tendril may operate in a dual fashion, acting as a grappling iron before commencing to twine, frequently the extreme distal ends of such tendrils, which may be branched, form very small recurved hooks (**61a**) or occasionally suckers (**229b**). Once a tendril has become anchored at any point, the remaining portion may continue to twist resulting in a spring shape. Such twisting may be clockwise over one portion of the tendril and anticlockwise over the remainder. A leaf tendril will usually have a bud (or shoot) in its axil, a stem tendril (**122**) will be subtended by a leaf (or its scar). However, interpretation is not always easy; the tendril of the Cucurbitaceae, which appears to be a stem tendril in the axil of a leaf, may in fact represent the prophyll of the bud in the axil of that leaf (**122**).

Fig. 68b. *Mutisia retusa*
Each simple leaf terminates in a tendril.

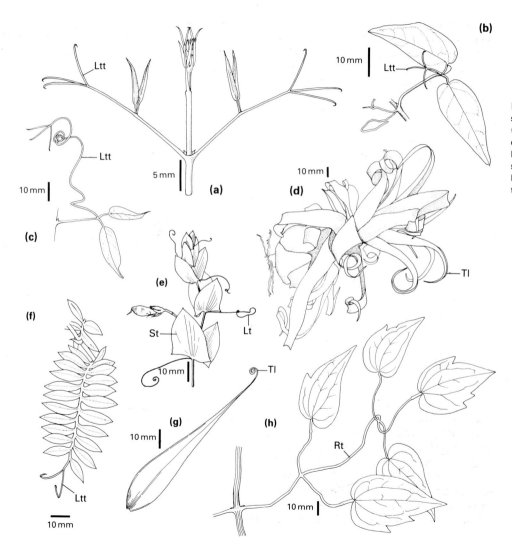

Fig. 69. a) *Bignonia* sp., shoot apex; b) *Bignonia ornata*, single leaf at node (second not shown); c) *Pyrostegia venusta*, single leaf; d) *Tillandsia streptophylla*, whole plant; e) *Lathyrus aphaca*, shoot apex; f) *Mutisia acuminata*, single leaf; g) *Littonia modesta*, single leaf; h) *Clematis montana*, single leaf at node (second not shown). Lt: leaf tendril. Ltt: leaflet tendril. Rt: rachis tendril. St: stipule. Tl: leaf tip tendril.

Fig. 70a, b. *Zombia antillarum*
a) The spine covered stem;
b) the sheath of each leaf is splayed out at the junction with the petiole into a fan of spines.

The whole or part of a leaf may be represented by a woody and more or less persistent spine (spine, thorn, prickle 76). A leaf spine can usually be recognized as such as it subtends a bud or shoot. Conversely, stem spines (124) will be in the axil of a leaf or leaf scar (6). However, care must be exercised as an apparent stem spine may in fact be formed from the first leaf or leaves of an axillary shoot (e.g. **71c, e, 203b**). The petiole only of the leaf may in whole (**40b**) or in part (**41a, b, c, d**) become woody and pointed after the detachment of the lamina or the leaf may bear stipular spines (**57f**). Occasionally a few leaflets only of a compound leaf develop as spines as in the case of climbing palms (**71f**). Alternatively the whole leaf (possibly including stipules if present) takes the form of a spine (**71c**) or spines (**71a**). In such cases the plant is dimorphic (30) having two distinct leaf types (i.e. spine and foliage leaf in this case) or all the leaves on the plant may occur as spines (most Cactaceae 202). A distinctive form of spine that is foliar in origin occurs in some palms (such as *Zombia* **70b**) in which the leaf sheath persists after the loss of the petiole and lamina and the veins in the distal portion of the remaining sheath form spines radiating out apparently from the trunk of the tree (**70a**).

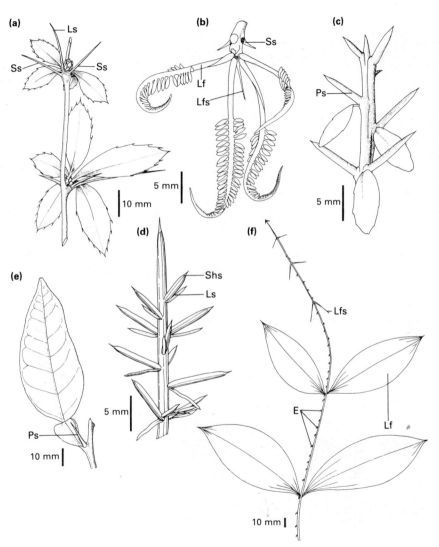

Fig. 71. a) *Berberis julianae*, portion of shoot; b) *Parkinsonia aculeata*, single young palmate/pinnate leaf; c) *Microcitrus australasica*, shoot apex; d) *Ulex europaeus*, shoot apex; e) *Citrus paradisi*, single leaf at node; f) *Desmoncus* sp., distal end of leaf. E: emergence. Lf: leaflet. Lfs: leaflet spine. Ls: leaf spine. Ps: prophyll spine (66). Shs: shoot spine (124). Ss: stipule spine (56).

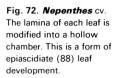

Fig. 72. **Nepenthes** cv. The lamina of each leaf is modified into a hollow chamber. This is a form of epiascidiate (88) leaf development.

The leaves of plants in a limited number of families (Droseraceae, Cephalotaceae, Lentibulariaceae, Nepenthaceae, Sarraceniaceae, and Dioncophyllaceae) form structures that trap insects and other similar-sized animals. Once caught the insect will be digested and absorbed over a period of time. The classical descriptions of insectivorous plants are to be found in Darwin (1875). Leaf traps are of two general types: sticky leaves (**73a, b, 36a, 81g**) with or without elaborate glandular tentacles, the leaves usually curling up to enclose caught insects; and epiascidiate (**88**) leaves, i.e. leaves forming a container into which the insect falls (**72, 31d, 89c, 43e**), flies, or is sucked (**73e**). The mode of development of pitcher-type leaves is described in section 86, these leaves frequently have deposits of loose wax flakes around the inner rim of the trap opening, which become stuck to insects' feet and speed the fall into the container. The epiascidiate leaf of *Utricularia* (**73e**) differs in that it is active in its action: the container has a lid which opens inwards in response to tactile stimulus of hairs at its entrance, and the structure of the bladder is such that water plus insect is instantaneously sucked in, water pressure being greater outside the trap than inside (Lloyd 1933). Rapid response to stimulation is also seen in *Dionaea muscipula* (**73f**) in which repeated pressure on hairs of the adaxial side of the leaf results in the two halves of the leaf snapping together. A complete account of insectivorous plants is given by Juniper *et al.* (1989).

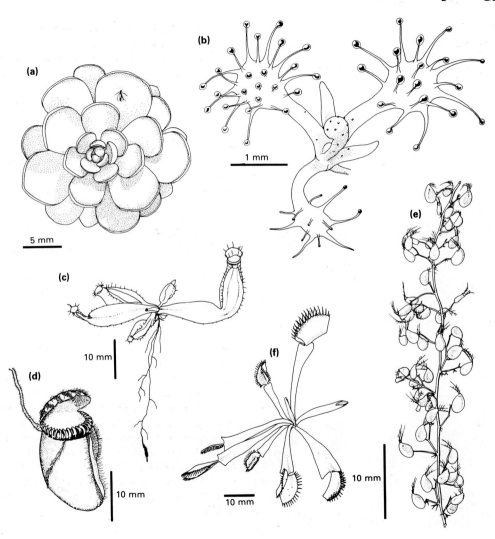

Fig. 73. a) *Pinguicula lanii*, leaf rosette from above; b) *Drosera capensis*, seedling; c) *Nepenthes khasiana*, seedling; d) *Cephalotus follicularis*, single leaf; e) *Utricularia minor*, portion of shoot; f) *Dionaea muscipula*, seedling.

It is conventional to interpret flowering plant structure in terms of four categories of organ—leaf, stem, root, and trichome (206). However, in many instances strict rigid adherement to this scheme creates major problems (206–212) or conflicting opinion of interpretation (4, 122). Classically, a leaf is expected to be a determinate (90) lateral appendage on a stem and not itself to bear other leaves or stems. Nevertheless, epiphylly (growth on a leaf) is not uncommon (Dickinson 1978). Many species are to be found which bear inflorescences or vegetative buds located on leaves in a variety of positions (75h–n). Such an occurrence will usually be a regular and normal feature for the given species, regardless of its apparent inconsistency with conventional morphological 'rules' although a range of epiphyllous structures arise in response to attack by mites in some plants (Ming *et al.* 1988). Conventionally a bud, be it potentially an inflorescence or a vegetative structure, is expected to be located in the axil of the leaf (4), not 'carried up' and positioned out on the leaf petiole or blade. There are a number of ways in which an epiphyllous structure may develop. One theoretical explanation is that the axillary bud has become fused (post-genital fusion) on to its subtending leaf after the independent growth of both. This is rarely observed. A second developmental explanation involves ontogenetic displacement. In the earliest stages of growth, cells below both the young bud primordium and subjacent leaf primordium divide actively and the bud and leaf grow out as one unit, i.e. they never have a separate existence. This sequence of

intercalary growth undoubtedly occurs in many instances and will result in an inflorescence or vegetative bud apparently sitting on a leaf and qualifying for the traditional explanation of adnation (**74**).

Epiphylly can result from a second developmental phenomenon with or without the occurrence of ontogenetic displacement. One (or more) area of cells on a leaf primordium retains its meristematic ability, initially common to all the cells of the primordium, and subsequently becomes organized into an independent shoot system. This is referred to as heterotopy. A good

example is that of *Streptocarpus* (208). Such heterotopies ('other place') are in direct conflict with the classical interpretation of plant growth which will have to dismiss them as 'adventitious' structures (98, 178, 232). Nevertheless, heterotopy is well documented and produces in conjunction with ontogenetic displacement, inflorescences on leaves (**75h–n**), leaves on leaves (Maier and Sattler 1977), vegetative detachable buds with roots on leaves (**75a, c, e**) and even apparently embryo-like structures on leaves (Taylor 1967). During the development of leaves of *Bryophyllum* species, patches of meristematic

Fig. 74. *Spathicarpa sagittifolia*
A row of flowers, representing an inflorescence spike (**141c**), remains attached to the subtending bract (or spathe) during development, as in Fig. **75k**.

cells stop dividing at intervals along the leaf, giving the leaf at first an indented margin. Subsequently these heterotopic areas recommence development to produce the detachable buds (**233**, cf. **227**). A number of other apparently extraneous structures may be found on the leaves. These include adventitious roots (**98**), galls (**278**), glands (**80**), food bodies (**78**), emergences (**76**), and stipels (**58**). Recognition of an epiphyllous structure is not always clear; the *Pleurothallis* sp. shown here (**75d**) has a conventional morphology with a terminal inflorescence located very close to the distal foliage leaf, i.e. it is not an example of epiphylly.

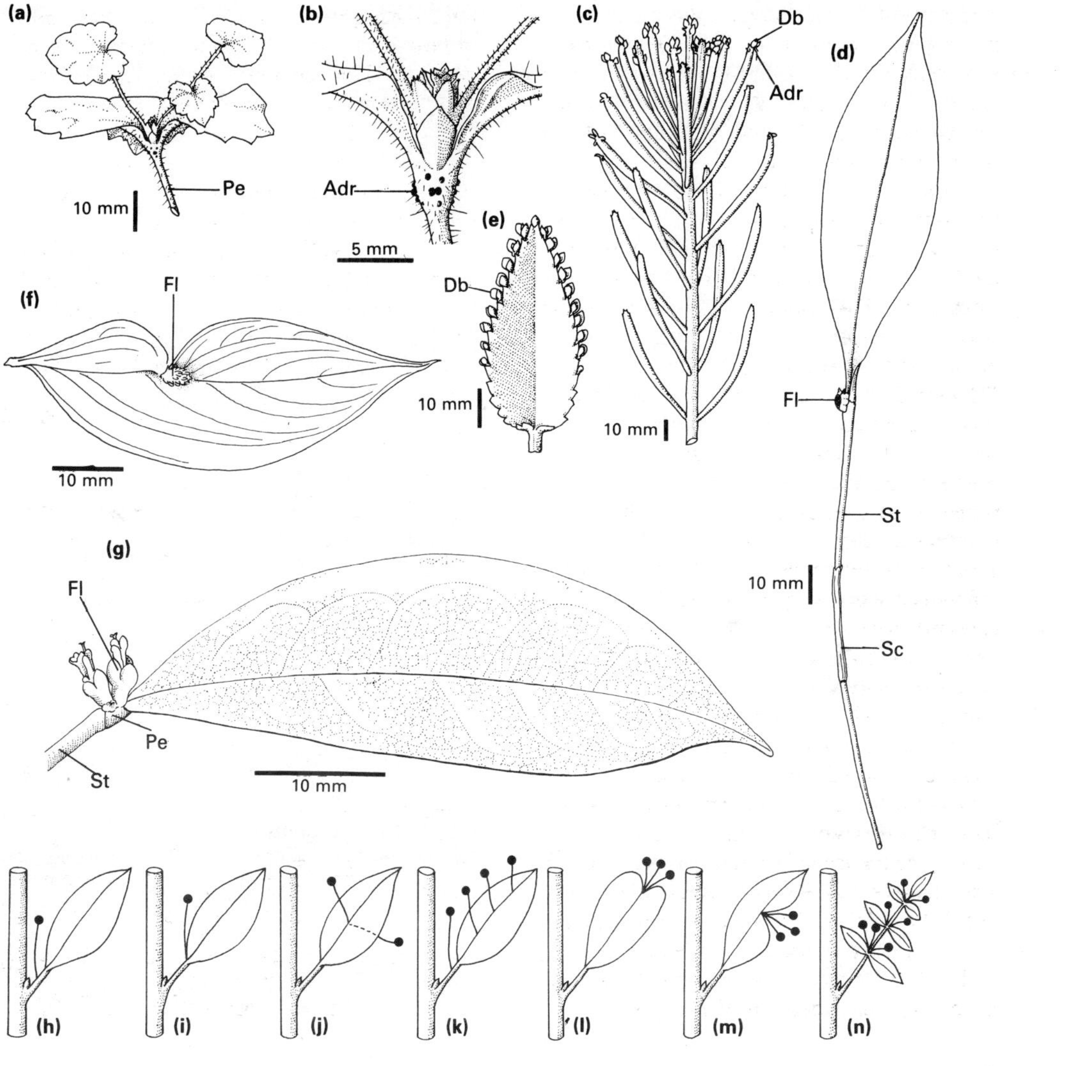

Fig. 75. a, b) *Tolmiea menziesii*, single leaf and close view lamina/petiole junction; c) *Bryophyllum tubiflorum*, end of shoot (cf. **227**); d) *Pleurothallis* sp., end of shoot (apparent epiphylly only); e) *Bryophyllum diagremontanum*, single leaf (cf. **233**); f) *Polycardia* sp., single leaf; g) *Tapura guianensis*, single leaf; h)–n) epiphyllous locations, after Dickinson (1978). Adr: adventitious root. Db: detachable bud. Fl: flower(s) Pe: petiole. Sc: scale leaf. St: stem.

Spiny structures are quite common features of above-ground plant parts. The terminology associated with these features is not consistently utilized, the terms spine, prickle, and thorn, being found more or less interchangeably. In this book 'thorn' is not employed; a 'spine' represents a modified leaf (leaf spine 70), stipule (56), stem (stem spine 124), or root (root spine 106), and 'prickle' is applied to sharp usually woody structures that develop from a combination of the epidermis of an organ plus subepidermal tissue (and also sharp structures on a leaf edge 7). The general term applied to a structure with this epidermal/subepidermal origin is emergence (stem emergence 116, leaf emergence 77), the consistent feature of an emergence being that it will not develop in the expected location of a leaf or shoot primordium (cf. phyllotaxis 218), representing as it does an additional form of organ. Leaf emergences vary considerably in size and shape and may be confined to the leaf margin or to either the upper (adaxial) or lower (abaxial) surface of the lamina or petiole. In compound leaves emergences may be found on the rachis between adjacent leaflets (77d). Emergences are not always haphazard in their location, in *Acacia seyal* (117d) a prickle occurs very close to each stipule. The stipules themselves are ephemeral and soon drop leaving a small and easily overlooked scar. A casual glance would doubtless suggest that the plant has spiny stipules.

Fig. 76. *Solanum torvium*
Emergences on leaf surface. They are also present on the stem (116).

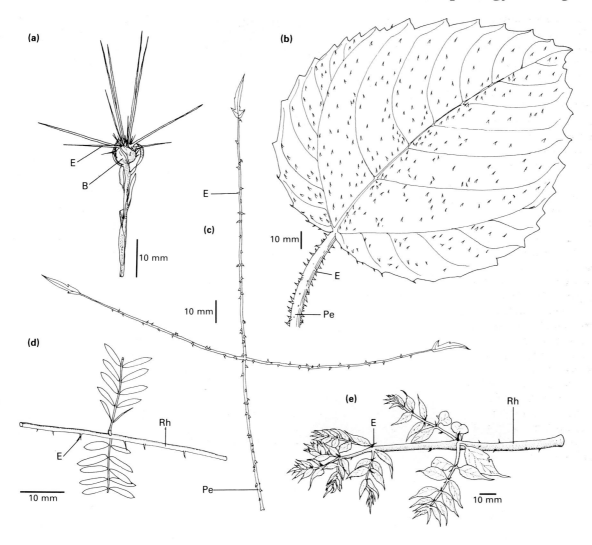

(a)

E

B

10 mm

(b)

10 mm

E

Pe

(c)

E

10 mm

(d)

E

Rh

10 mm

Pe

(e)

E

Rh

10 mm

Fig. 77. a) *Centaurea* sp., inflorescence; b) *Laportea* sp., single leaf; c) *Rubus australis*, single leaf; d) *Acacia* sp., portion of leaf; e) *Aralia spinosa*, portion of leaf. B; bract. E: emergence. Pe: petiole. Rh: rachis.

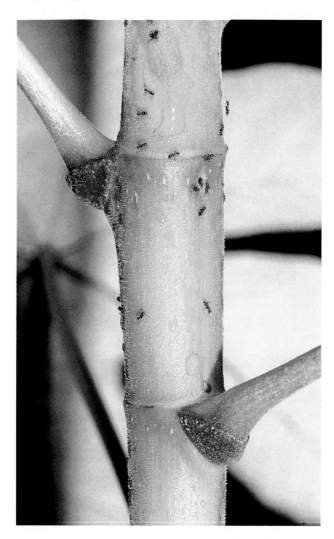

Fig. 78. *Cecropia obtusa*
On the abaxial side of each petiole base is a pad of tissue producing a constant supply of food bodies. On the opposite side of each node, just below the stipule scar, is a weak spot that is excavated by ants to provide an entrance to the hollow internode nesting site.

A range of structures commonly referred to as 'food bodies' occur on the surface of some plant leaves and anatomically can represent either trichomes (80) or emergences (76) which are secreting usually edible proteinaceous substances. Unfortunately, as each new example of this phenomenon has been discovered the food body has been given a specialist term. A number of such food bodies are listed here.

(1) Beltian bodies (after Belt)—these are food bodies occurring at the ends of leaflets in *Acacia* species (79);

(2) Müllerian bodies (after Müller)—food bodies born on a swelling (trichilium) at the base of leaf petioles of *Cecropia* species (78);

(3) Beccariian bodies (after Beccari)—found in various locations on leaf and stipule of *Macaranga*;

(4) Pearl bodies on *Ochroma* (on the leaves and stems);

(5) Food cells on *Piper* species found in domatia (204) in the petiole.

Similar structures usually secreting oil (oil bodies or elaiosomes) are found on the seeds of many plants and then usually act as ant attractants. Small structures found on the surface of leaves of many *Passiflora* species mimic butterfly eggs.

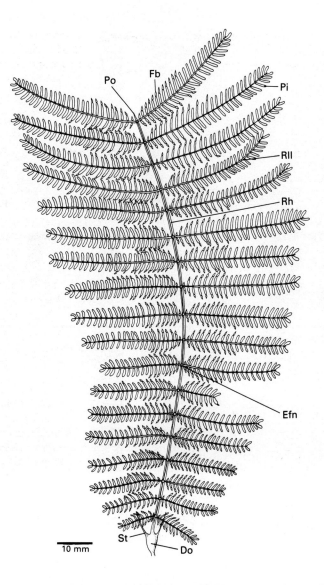

Fig. 79. *Acacia hindisii,* single leaf. Do: domatium (204). Efn: extra-floral nectary. Fb: food body. Pi: pinnule. Po: pointlet. Rh: rachis. Rll: rachilla. St: stipule.

Various structures are to be found developing on the surface of leaves, stems, and roots. These include galls (278), nodules (276), adventitious buds (232, 178), and epiphyllous organs (74). In addition many plant parts bear hairs (trichomes) which may be unicellular or multicellular and which are epidermal in origin, and emergences (76) which are usually more substantial and of epidermal plus subepidermal origin. The range of trichome anatomy is largely outside the scope of this book, but the more bulky glandular types can be very conspicuous. Some glands are undoubtedly of subepidermal origin and therefore are strictly emergences, but for convenience these are illustrated here. Emergences of a woody nature are described elsewhere (76, 116). Glandular structures may secrete salt (salt glands), or water (hydathodes), or sugar solutions (extra-floral nectaries **81d, e**). A review of the range of morphology and terminology associated with nectaries is given by Schmid (1988). Glands of many insectivorous plants (**36a, 73b, 81g**) secrete a very viscous substance. Solid secretions are referred to as food bodies (78). Two types of glandular trichome are associated with the protection or unfolding of buds. In just two superorders, the Alismatiflorae and the Ariflorae, glandular trichomes occur in the axils of vegetative leaves. These trichomes are termed squamules. Similar glandular trichomes are associated with the buds of many other plants where they are referred to as colleters.

Fig. 80a. *Passiflora glandulosa*
Ant feeding at extra-floral nectary on the surface of the flower bud.

Fig. 80b. *Acacia lebbek*
A cup shaped extra-floral nectary on the upper (adaxial) surface of the leaf petiole at the edge of the pulvinus (46). Dead stipules (52) about to fall.

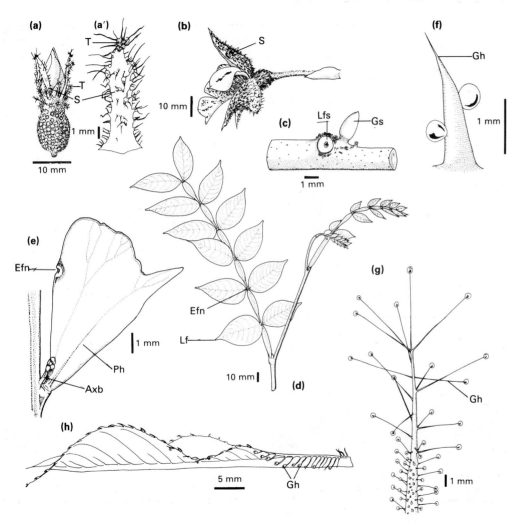

Fig. 81. a, a¹) *Osbeckia* sp., flower bud and single sepal; b) *Dendrobium finisterrae*, single flower; c) *Cassia floribunda*, portion of leaf rachis (cf. **59a**); d) *Inga* sp., end of shoot; e) *Acacia pravissima* leaf at node (cf **43d**); f) *Laportea* sp., single stinging hair (cf. **77b**); g) *Drosera binata*, leaf tip; h) *Impatiens sodenii*, single leaf. Axb: axillary bud. Efn: extra-floral nectary. Gh: glandular hair. Gs: glandular stipel. Lf: leaflet. Lfs: leaflet scar. Ph: phyllode. S: sepal. T: trichome.

Fig. 82. *Graptopetalum* sp.
The spirally arranged leaves of each
rosette are fat and fleshy.

Parts of plants are generally described as
'succulent' if they are particularly fleshy, not
woody, to the feel and noticeably watery if
squashed. Roots (**111**), stems (**203**), or leaves
can store water and are associated with
environments subjected to conditions of drought.
The leaf bases of bananas forming a pseudostem
(**50**) can be described as succulent, likewise the
thick scale leaves constituting a bulb (84). More
pronounced succulency is found in xerophytic
and epiphytic plants (potentially dry conditions)
and in halophytes (saline conditions). The fleshy
leaves of such plants may be bifacial (**83c**),
cylindrical (i.e. unifacial **83j**), or approximately
spherical in shape (**83a**). If internodes between
leaves are very short, then successive leaves will
be partly enveloped by older leaves. This is
particularly pronounced if the leaves are in
opposite decussate pairs (**83i**) and especially so if
each pair is united around the stem
(connate 234). 'Stone plants' (e.g. *Lithops* spp.)
take this form (**83b**).

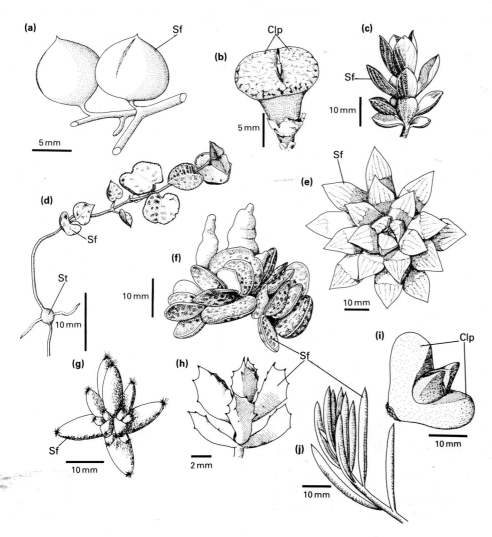

Fig. 83. a) *Senecio rowleyanus*, two leaves; b) *Conophytum mundum*, leaf pair; c) *Coleus caerulescens*, end of shoot; d) *Ceropegia woodii*, portion of shoot; e) *Haworthia turgida* ssp. *subtuberculata*, leaf rosette from above; f) *Adromischus trigynus*, leaf rosette; g) *Trichodiadema densum*, leaf rosette from above; h) *Oscularia deltoides*, end of shoot; i) *Cheridopsis pillansii*, leaf pair; j) *Othonna carnosa*, end of shoot. Clp: connate (234) leaf pair. Sf: single succulent leaf. St: stem tuber (138).

Fig. 84. *Urginea* sp.
The base of each leaf sheath is swollen, the whole forming a bulb. In the axil of each leaf is a vertical row of accessory buds (236) developing as detachable bulbets.

A bulb consists of a short, usually vertical, stem axis bearing a variable number of fleshy scale leaves. Its organization is usually imprecise in dicotyledons but more precise in monocotyledons and has acquired a considerable descriptive terminology. The outer scale leaves of the bulb may be membraneous rather than fleshy. They may develop in this way or represent the collapsed remains of a previous season's fleshy leaves. Internodes between leaves rarely elongate and adventitious roots develop from the basal part of the stem (or 'stem plate'); these roots are often contractile (**107e**). The bulb may produce inflorescences in the leaf axils in which case the monopodial main stem axis can bear a series of bulb-like structures (**85d**), the successive stem plates possibly remaining after the leaves have decayed. Alternatively, the inflorescence can be terminal in which case one or more axillary buds will develop as renewal (replacement or regenerative) daughter bulbs producing a sympodial series. Additional (increase or proliferative) bulbs smaller in size (bulbets) than the main renewal bulbs may be present, and form a mode of vegetative multiplication (170, 172). Green foliage leaves will develop usually at the distal end of the bulb axis, alternatively the bulb is constructed of fleshy leaf bases, each leaf then having a photosynthetic, temporary lamina. Loosely organized bulbs are typical of dicotyledonous plants.

The majority of monocotyledonous bulbs have a more compact structure resulting from the concentric insertion of the leaves on the short stem plate, and the sequence of parts in a bulb can be precise (**85e**). For example a fixed number of concentric protective (i.e. membraneous and/or somewhat woody) scale leaves at the proximal end, being followed by a fixed number of fleshy storage leaves, possibly only one, in turn followed by a fixed number of foliage leaves. In *Hippeastrum*, the bulb is constructed sympodially and each sympodial unit bears four leaves and a terminal inflorescence. Axillary buds may be formed subtended by some or all leaf types and will develop into new bulbs or inflorescences if the inflorescence is not terminal. An axillary bulb may be physically displaced away from its parent bulb at the end of an elongating stolon (cf. dropper 174). The bulb of garlic (*Allium sativum* **85b**) consists of a proximal series of membraneous protective scale leaves subtending no buds, a series of membraneous scale leaves each subtending a number of axillary buds, the 'cloves' (accessory buds **84, 236**) and most distally a number of foliage leaves. The axis terminates in a sterile inflorescence. Each clove has an outer protective prophyll (**85b**), the second leaf is a storage scale leaf, the third leaf is a foliage leaf with little lamina, and subsequent leaves will be fully functional foliage leaves. It is customary and useful to display the construction of a bulb by means of a conventionalized 'exploded' diagram in which the internodes are elongated with successive leaves drawn as a nest of inverted cones (**85d, e**). The commercial importance of bulbs of various kinds has led to a wide range of terminology to describe their various features. The applied aspects of bulb construction can be consulted in Rees (1972).

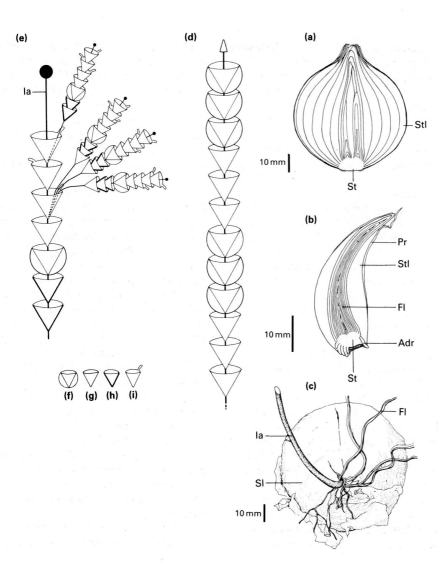

Fig. 85. a) *Allium cepa*, longitudinal section entire bulb; b) *Allium sativum*, longitudinal section single axillary bud, clove, of bulb; c) *Bowiea volubilis*, whole bulb; d) diagram of construction of typical monopodial bulb, e) of sympodial bulb, f) storage scale leaf; g) membranous scale leaf, h) protective scale leaf, i) foliage leaf. Adr: adventitious root. Fl: foliage leaf (yet to extend in b). Ia: inflorescence axis. Pr: prophyll. Sl: scale leaf. St: stem. Stl: storage leaf.

Active cell division and enlargement in the various meristems of a leaf primordium (18) can result in a leaf of virtually any shape. A typical dorsiventrally flattened leaf (bifacial leaf **87f**) with a 'top' (adaxial) side and a 'bottom' (abaxial) side results if the meristems along the edge of the leaf primordium are active. Increase in the number of cells at the centre of the adaxial side of the leaf (adaxial meristem **19d**) will give rise to the thickening of the midrib. In some leaves where an adaxial meristem activity is marked, lateral extension is suppressed resulting in a more or less cylindrical leaf; such a leaf is termed unifacial as it is radially symmetrical (Kaplan 1973b) and does not have the two sides of a bifacial leaf. The unifacial leaf may remain cylindrical (terete or centric **87g**) or subsequently become flattened bilaterally (isobilateral or ensiform **87h**). The phyllodes of *Acacia* and other plants are formed in this way (42). The base of an ensiform leaf retains a bifacial form which is usually folded (conduplicate **37j**) and the bases of successive leaves demonstrate equitant vernation (**39g**). The leaf of *Dianella* has a conduplicate base, an ensiform middle portion, and a bifacial distal end (**51a**). Terete (cylindrical) leaves result from the development of the upper leaf zone of the leaf primordium in both monocotyledons and dicotyledons and are therefore homologous (20). Localized subtleties of meristematic activity also give rise to peltate and asciidiate leaves in some species (88).

Fig. 86. *Tillandsia usneoides*
The adult plant has no roots, atmospheric water being absorbed by the fine terete leaves.

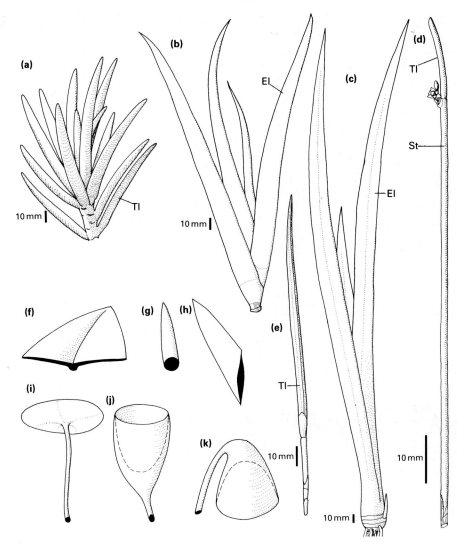

Fig. 87. a) *Senecio* sp., end of shoot; b) *Oberonia* sp., end of shoot; c) *Iris pseudacorus*, foliage leaves at distal end of rhizome; d) *Ceratostylis* sp., stem with distal terete leaf; e) *Reichenbachanthus* sp., stem with distal terete leaf, f) bifacial, g) terete, h) ensiform, i) peltate, j) epiascidiate, k) hypoascidiate. El: ensiform leaf. St: stem. Tl: terete leaf.

Activity of the various areas of meristematic cells present in a developing leaf primordium (18) commonly gives rise to a bifacial leaf with an upper (**87f**) surface (ventral, adaxial) and a lower surface (dorsal, abaxial). However, leaves flattened in the ventral plane (ensiform **87h**) and cylindrical leaves (terete **87g**) are not uncommon. Similarly, differential meristematic activity can give rise to a peltate leaf (**87i**) in which a more or less circular lamina has the petiole attached near the centre (**36b, 89b, d**). This shape can also occur as a teratology (peltation 270) in any leaf, particularly one that normally has basal lobes. The lamina of a peltate leaf is flat or slightly dished; if meristematic activity continues, the lamina can become funnel-shaped forming a container and the leaf is termed ascidiate. Normally the inside of the container is developmentally equivalent to the top of a peltate leaf, and the outer surface equivalent to the underside of a peltate leaf (epiascidiate). The distinctive leaf of a pitcher plant conforms to this arrangement (**89c**). The epiphyte *Dischidia* has two forms of leaf, bifacial on a climbing stem and ascidiate leaves developing near the branch of the supporting tree. Adventitious roots (98) grow into the opening of the ascidiate leaf which contains debris (**89f**). Bladder leaves of the Lentibulariaceae are ascidiate (**73e**), and are variously developed from highly dissected submerged leaves of these water plants which have no roots (cf. **91e**). Very rarely an ascidiate leaf results from the development of a pouch in which the lower surface is inside—a hypoascidiate leaf. Bracts (62) subtending flowers of *Pelargonium* can take this form, as do those of *Norantea* (**88a, b**, and frontispiece).

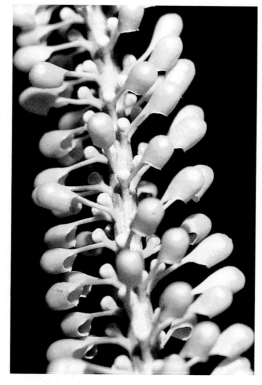

Fig. 88a, b. *Norantea guyanensis*
The bract (62) subtending each flower is hypoascidiate initially developing as an inverted spoon shape (a) and then forming a hollow chamber (b) containing extra-floral nectaries (80). The final form is shown in the frontispiece.

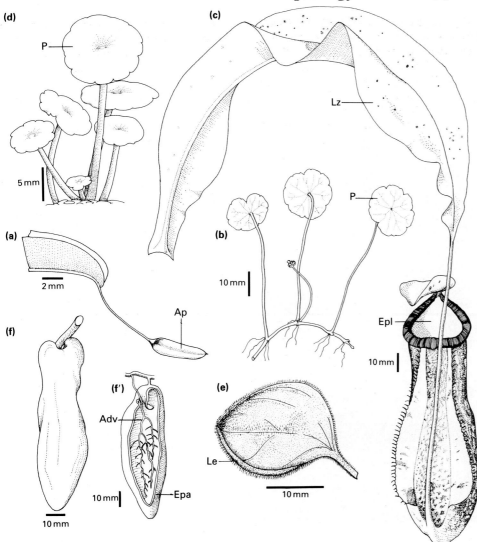

Fig. 89. a) *Cassia floribunda*, abnormal leaf tip; b) *Hydrocotyle vulgaris*, stolon bearing leaves; c) *Nepenthes × coccinea*, single leaf; d) *Umbilicus rupestris*; e) *Justicia suberecta*, single leaf; f, f¹) *Dischidia rafflesiana*, single leaf and section of leaf. Adv: adventitious root. Ap: abnormal peltate development (peltation). Epa: epiascidiate leaf. Epl: epiascidiate lamina (upper leaf zone 20). Le: inrolled leaf edge (not peltation). Lz: lower leaf zone. P: peltate leaf.(f¹ after Massart 1921).

Fig. 90. *Guarea glabra*
A young tree. Each apparently woody slender stem bearing simple leaves is in fact a long-lived growing compound leaf (**91f**).

A leaf, particularly on a woody plant, is generally found to be a temporary structure, developing relatively rapidly to a finite size (i.e. it is determinate) and persisting until dislodged by drought or frost or loss of vascular connection on an expanding stem axis (48). A branch system is seen to be more permanent. However, twigs and branches are often shed (268) and conversely some plants possess leaves that grow progressively for some time (i.e. they are more or less indeterminate). This results from a proximal intercalary meristem in the Gramineae (180) and other monocotyledons. In some dicotyledons, the distal end of a pinnate leaf retains its capabilities for cell division and the final length of the leaf is attained over an extended period by the periodic production of extra pairs of leaflets (**90, 91f**). Such structures, delayed in their appearance, can be preformed, i.e. the whole leaf develops initially but its parts mature in sequence (**91a, b**) from leaf base to leaf apex and the leaf is thus strictly speaking determinate. Alternatively, the leaf is truly indeterminate and the apical meristem of the leaf continues to function, initiating new growth periodically for several years as in *Guarea* (Steingraeber and Fisher 1986), (epigenesis **91c, d**). The oldest, i.e. proximal leaflets, fall off in the meantime and the leaf rachis increases in girth due to cambial activity (such cambial activity is sometimes also found in the petiole of other long-lived but determinate leaves **40a**). Indeterminate leaves often bear inflorescence primordia in association with the new leaflet primordia (epiphylly 74). The underwater leaves of *Utricularia* (cf. 206), are indeterminate in development and form an apparent much branched structure (**91e**). The unique phyllomorph (208) of some *Streptocarpus* spp. behaves in the manner of an indeterminate simple leaf.

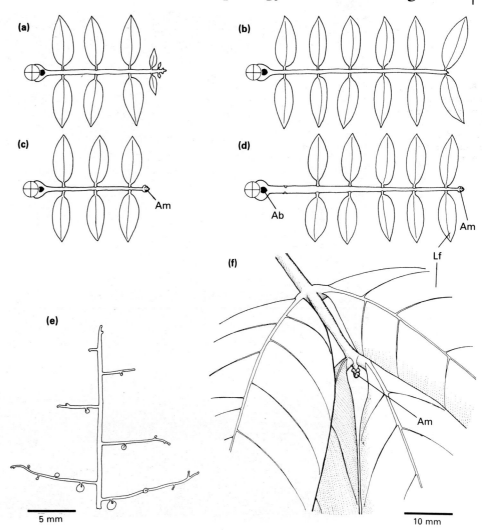

Fig. 91. a, b) determinate leaf developing over a long time interval from preformed leaflets; c, d) indeterminate leaf developing new leaflets from an apical meristem; e) *Utricularia reniformis*, end of indeterminate leaf; f) *Guarea glabra*, distal end of compound leaf (**90**). Ab: axillary bud. Am: apical meristem (of the leaf). Lf: leaflet.

Fig. 92a. *Jubaea spectabilis*
Reduplicate attachment of leaflets to midrib of leaf.

Fig. 92b. *Phoenix dactylifera*
Induplicate attachment of leaflets to midrib of leaf.

The leaves of the palms (Palmae) show a sufficient number of distinctive morphological features to warrant separate description. All palm leaves have a lamina, a petiole, and a sheath, the lamina being mostly of three general shapes—palmate (**93a**) lacking a rachis, pinnate in which leaflets are born on the rachis (**93c**), and costapalmate, an intermediate shape in which palmately arranged leaflets are born on a very short rachis or costa (**93b**). (A few palms have simple leaves; *Caryota* has a bipinnate leaf **93d**.) The most distinctive feature of the palm leaf occurs in the development of the leaflets (Dengler *et al.*, 1982, Kaplan *et al.*, 1982a, b). These do not arise by differential growth rates in meristems along the leaf primordium edge (18). Instead, differential growth in the expanding leaf lamina causes the lamina to become plicate (**37i**), i.e. folded into ridges and furrows. There is then a subsequent separation of rows of cells between plications giving rise to the distinct leaflets. Strips of dead cells occur at the edges of palm leaves and are known as reins, or lorae; they form a conspicuous feature of some palms (**93d**). In palmate and costapalmate leaves the splitting may not extend all the way from the lamina edge to the centre; this is a specific variation. One effect of the splitting between plications of a palm leaf is that the attachment of an individual leaflet, or 'finger', to the rachis or petiole can take two forms. It may be reduplicate (**92a**) or induplicate (**92b**). Almost all 'fan' leaves (palmate and costapalmate) are induplicate; most 'feather' leaves (pinnate) are reduplicate and have a terminal pair of leaflets (paripinnate **23e**).

The few that are induplicate are imparipinnate (**57f**) having a single terminal leaflet. A ridge of tissue, the hastula (**93a′**), is present at the junction of petiole and lamina in some palmate and costapalmate leaves. It may be on the adaxial side, the abaxial side, or both. (A similar structure occurs on leaves in the Cyclanthaceae.) The sheaths of palm leaves may persist on the tree for many years in the form of a fibrous mat (**51b**), or as stumps, splitting in the mid line due to stem expansion, or forming a collection of spines (**70a, b**), the spines representing the fibrous vascular bundles of a ligule at the junction of sheath and petiole. Non-spiny ligules occur in a number of palms. Spines also occur in the form of modified adventitious roots (106), as spines on long thin modified inflorescences (flagellum), or as emergences on leaf (**71f**) or stem. The leaves of rattans (climbing palms) often bear distal pairs of leaflets modified into spines or reflexed hooks on an extended rachis or 'cirrus' (**71f**). A full account of the morphology of palms is given by Tomlinson (1990).

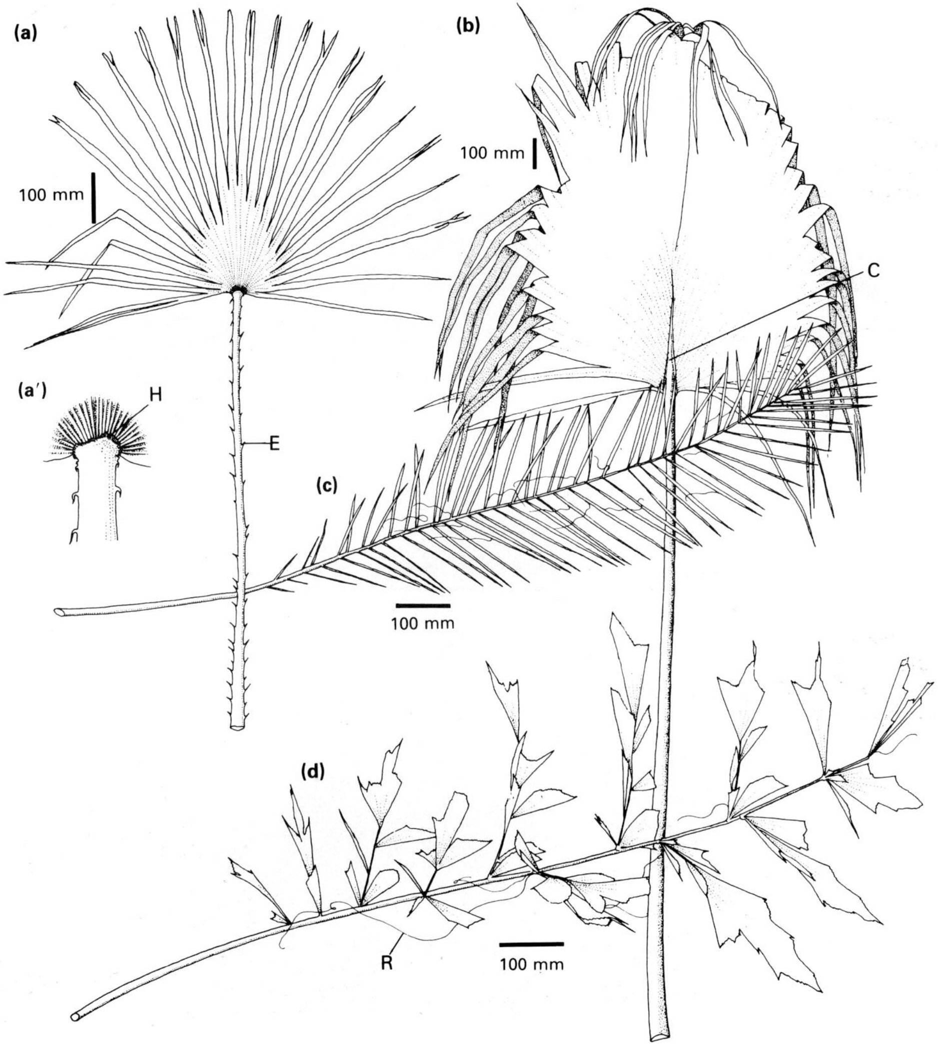

Fig. 93. a, a¹) *Livistonia* sp., single palmate leaf and close view of lamina/petiole junction; b) *Sabal palmetto*, single costapalmate leaf; c) *Phoenix dactylifera*, single pinnate leaf; d) *Caryota* sp., single bipinnate leaf. C: costa. E: emergence (76). H: hastula. R: reins.

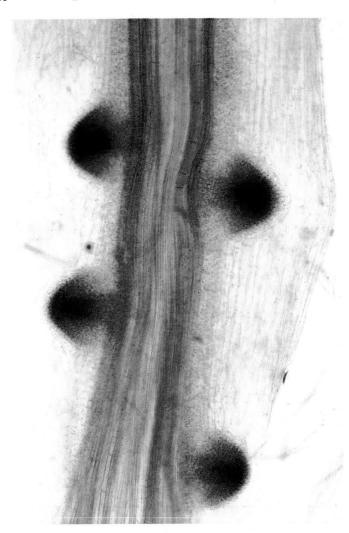

Fig. 94. *Pisum sativum*
A portion of root rendered semi-transparent showing the internal (endogenous) location of lateral root primordia.

A root develops from a root primordium, a group of meristematic cells originating below the surface of an existing root or shoot (endogenous development—produced within **94**). The first root of the embryo and all subsequent roots increase in length due to cell division and enlargement behind the root apex. The region of the root apical meristem is protected by a permanent covering of mature cells, the root cap, which is particularly obvious in some aerial roots (**95**). The root cap can be replaced by the root apex if the cap is damaged. Apart from the root cap, a root apex does not bear any other structures and thus contrasts with the shoot apex (112) which bears leaf primordia and associated axillary buds on its surface (exogenous development—borne externally), the shoot apex being protected by its enveloping leaves or other means (264). Some distance back from the root cap and apex a root may bear lateral roots. These lateral roots commence development from meristematic areas, with root primordia occurring beneath the surface of the parent root pushing their way out through the parent root cortex. In addition to lateral roots, other structures may develop on a root away from its apex: nodules in association with bacteria (276), mycorrhiza in association with fungi (276), and root buds (i.e. shoot buds on roots 178) capable of developing into new complete shoot systems. Elaborate root systems can develop in two basic ways. The initial radicle (162) of the seedling will bear many lateral root primordia possibly in some orderly sequence (96). The lateral roots can subsequently branch, and root cross-sectional area will increase due to cambial activity (16) as new lateral roots are added. Secondly, root primordia can arise endogenously in stem tissue giving rise to an extensive adventitious root system (98). Such roots are often associated with the nodes on the stem. This type of system is found in the majority of monocotyledons, the roots being incapable of extensive enlargement in girth. Adventitious roots in the Bromeliaceae can extend some distance in the stem cortex, growing parallel to the stem surface before finally emerging (intercauline roots). Root primordia present in an embryo before germination are referred to as seminal roots (162).

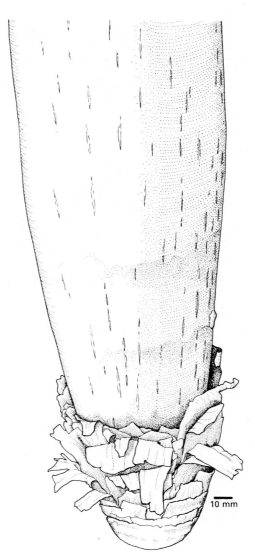

Fig. 95. *Pandanus nobilis*. Tip of aerial prop root (cf. **103**) showing massive root cap.

10 mm

Root systems are generally recognized to be of two basic types. In the first type the whole system is derived by growth and lateral branching of the seedling radicle (162) and is termed a primary root system; this type is typically found in dicotyledonous plants. In the second type, the primary root system is supplanted by an adventitious root system and is ubiquitous in the monocotyledons. An adventitious root develops from a root primordium arising in a stem or leaf (98). (The term adventitious is also occasionally applied to roots developing late and out of sequence in a primary root system.) Some dicotyledonous plants possess both types. Attempts to describe the varieties of branching of primary root systems take three approaches: a description of the overall form of the branching system, an investigation of the location of lateral root primordia in the developing root system, and analysis of the branching system in terms of branch orders (284), geometry and topology (mathematical description of branching) (Fitter 1982). An example of the type of classification that can be applied to primary root systems is given in Fig. **97a–f** modified from Cannon (1949). This system relies on the clear distinction of the vertical growth of the primary root and the various configurations of the first order lateral roots. First order laterals will bear second order laterals and so on. Four additional categories are applied to adventitious root systems (**97g–j**). Similar types of classification exist for tree root systems (100). Whatever the form of rooting, the details of the branching pattern depend on the

Fig. 96. *Bignonia ornata*
A climbing plant in which the central leaflet of each trifoliate leaf is in the form of a three pronged hook (cf. Fig. **69b**). There is a bud visible in the axil of each leaf. Also developing at this node is a pair of branched adventitious roots (98) visible just above each leaflet 'claw'.

location of the lateral root primordia. Root primordia result from the meristematic activity of patches of cells beneath the surface of existing roots (94). The sitings of primordia are not haphazard and varying degrees of orderliness can therefore be observed in the location of lateral roots ('rhizotaxis'). Lateral root primordia are frequently initiated in longitudinal rows within the parent root, the position of rows being governed by the arrangement of the vascular tissue in the centre of the root. Rows vary in number from two, three (**97k**), and four to many. The greater the number of rows, the less precise lateral root initiation appears to be. There can also be a degree of regularity of primordium spacing along any one row. (Mallory *et al.*, 1970).

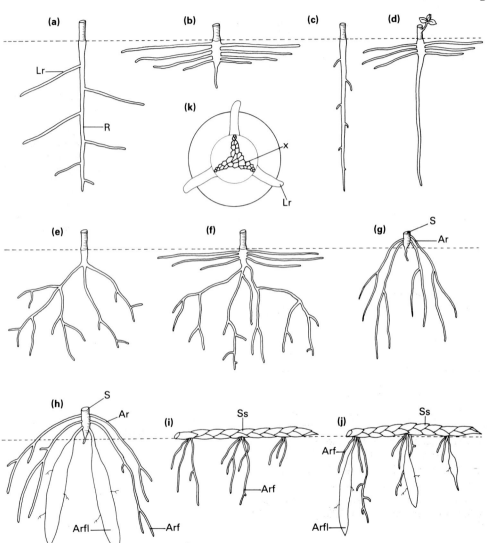

Fig. 97. Adapted from Cannon (1949). a)–f) Variations of primary root systems (lateral roots developing on radicle); g)–j) types of adventitious root systems. Roots developing on vertical (g, h) or horizontal (i, j) stem. k) Section through root having three-rowed xylem arrangement. Ar: adventitious root. Arf: fibrous adventitious root. Arfl: fleshy adventitious root. Lr: lateral root. R: radicle. S: stem. Ss: stem scale leaf. X: xylem tissue.

Fig. 98. *Philodendron* sp.
A number of adventitious roots develop at each node of the climbing stem. Some grow vertically downwards, others grow horizontally, wrapping around the support (the supporting plant demonstrates stem emergences, cf. Fig. **117c**).

Adventitious is an unfortunate adjective that literally means 'arriving from outside' and in morphology can be applied to any organ that is found in an atypical position. This is possibly appropriate in the case of an adventitious bud (232) occurring on the lamina of a leaf (74) because the vast majority of buds occur in the axils of leaves (4). Even then it is not necessarily unusual for the plant in question. The term is even more inappropriate in the case of adventitious roots, where it is applied to roots developing on stems or leaves, i.e. not forming part of the primary root system (96). In practically all monocotyledons the primary root system is short-lived and the whole functional root system of the plant is adventitious, the roots arising on the stem near ground level or below. This is particularly obvious in the case of rhizomatous monocotyledons (130). Similar elaborate adventitious root systems develop as a matter of course in many dicotyledonous plants having a rhizomatous or stoloniferous habit (132). In both cases, adventitious roots tend to be associated with nodes (i.e. they may be termed nodal roots) and the exact positions of development of the endogenous root primordia are governed by features of the vascular tissue at that node. This can result in quite precise patterns of nodal root arrangement, particularly in dicotyledons (96). Conversely, adventitious roots of climbing plants are often borne between nodes (99a).

Adventitious root primordia may be formed in the meristematic region of shoot apices and then develop into roots immediately or possibly much later when the supporting organ is old, or they may arise in old tissue by dedifferentiation, i.e. the return to meristematic activity, of selected patches of cells. The development of these new or latent primordia in an existing primary root system gives rise to an additional root system to which the term adventitious is sometimes applied, particularly in the case of tree roots. Thus the term adventitious root can be found applied either to a root 'out of place', i.e. on stem or leaf, or a root developing from old organs including old roots. Adventitious roots on stems are not always of one type. The classification of adventitious root systems of Cannon (1949) emphasizes this point. For example roots may be long, thin and anchoring, or much branched and fibrous (235a), or grow vertically upward or vertically downward (98). Root primordia can thus have a specific fate (topophysis 242) in some plants. Stems of *Theobroma* (cocoa tree) do not produce adventitious roots unless severed and allowed to form rooted cuttings; adventitious roots on 'chupon' stems (which grow vertically upward) themselves develop vertically downward, conversely adventitious roots developing on 'jorchette' stems, which grow horizontally, also develop horizontally. An extended account of adventitious roots, and others, is given in Barlow (1986).

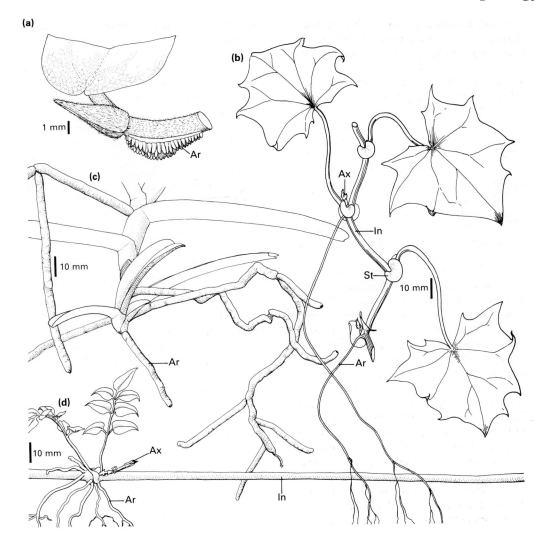

(a)

1 mm

Ar

(b)

Ax

In

St

10 mm

Ar

(c)

10 mm

Ar

In

(d)

10 mm

Ax

Ar

Fig. 99. a) *Ficus pumila*, end of climbing shoot (cf. **243**); b) *Senecio mikanoides*; c) *Acampe* sp., roots emerging from between leaf sheaths; d) *Jasminium polyanthum*, portion of trailing stem. Ar: adventitious root. Ax: axillary shoot. In: internode. St: stipule (52).

The branching systems of tree roots are exceedingly diverse, the architecture of the system of an individual tree changing considerably as it develops. Relatively young trees may have a tap root system based on the development of the radicle. Krasilnikov (1968) describes a range of variations of this theme (**101a, d–f**);which can be compared with root system descriptions of Cannon (1949) (**97**). This primary system can then become augmented or completely replaced by a secondary root system. The secondary system (sometimes referred to as adventitious 98) develops by the activity of root primordia on the old primary root system and the production of adventitious roots from stem tissue (**100**). A further distinction in a tree root system can be made between the skeletal system, i.e. the main framework which will be primary and/or secondary, and additional sub-systems of primary or secondary roots not contributing to the main supporting architecture. Additional distinctive features may be apparent such as buttresses (**101c, d**), stilt and prop roots (102), and pneumataphores (104). The roots of one individual tree can become naturally grafted to each other and such grafting has been recorded between the roots of neighbouring trees usually of the same species but occasionally of different species. The general phenomena of tree root architecture discussed here are those identified by Jenik (1978) in a tentative classification of tropical tree root systems (**101**) in which the primary root system is always more or less obliterated.

Fig. 100. *Pandanus* sp.
An elaborate prop root (102) formation in the manner of Fig. **101g**.

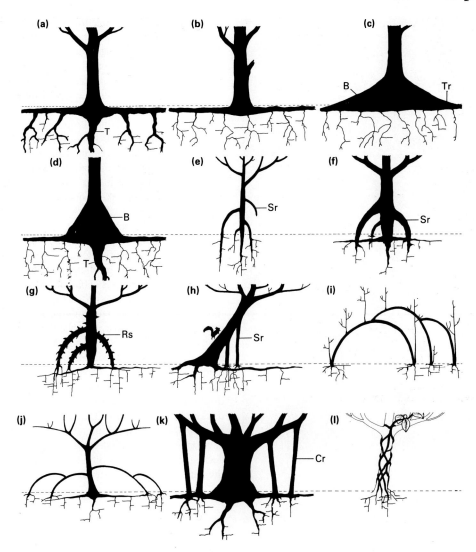

Fig. 101. Adapted from Jenik (1978). Tropical tree root systems. B: buttress. Cr: columnar root. Rs: root spine. Sr: stilt (prop) root. T: tap root. Tr: tabular root. (see 102).

Fig. 102. *Euterpe oleracea*
Prop roots on a palm. The small
outgrowths on the surface of each root
are pneumatorhizae (104).

Prop or stilt roots are adventitious roots (98) developing on the trunk or branches of a tree or the stem of a vertically growing herb. In a few exceptional cases, horizontal rhizome systems are supported as much as a metre above the ground on stilts roots (*Hornstedtia*, *Geostachys*, and *Scaphochlamys* in the Zingiberaceae and *Eugeissonia minor*, a palm). Prop roots are also found supporting pneumatophore roots (104). The tentative tropical tree root classification of Jenik (1978) includes a number of permutations of tree stilt root construction (**101e–k**). Stilt roots may themselves bear stilt roots (**101j**), Fig. **101i** indicates a similar result developing in this case by arching and rooting of shoot systems. Prop roots can take the form initially of spines (**101g**), which may subsequently elongate to form spine roots. Prop roots may be bilaterally flattened, forming flying buttresses (**101c, d**); these root buttresses may be positioned at the base of the tree trunk or form flattened tabular root structures running away at soil level from the tree. A prop root usually branches freely once it reaches the ground. It will retain its initial diameter in a monocotyledonous plant; in dicotyledonous plants it may remain very thin until rooted at its distal end and subsequently thicken into a prop or columnar root (**101k**). In many epiphytic plants long adventitious roots develop and grow into detritus or hang free as aerial roots. Such roots may anastomose around the supporting plant—'strangling roots' (**101l**).

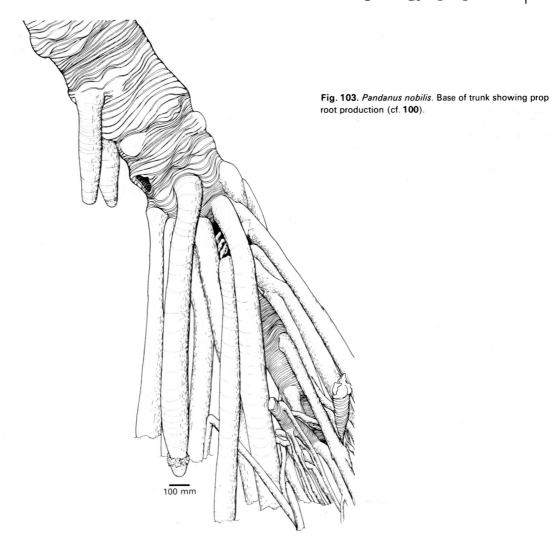

Fig. 103. *Pandanus nobilis.* Base of trunk showing prop root production (cf. **100**).

100 mm

Many woody plants living in swampy or tidal conditions show modifications of that part of the root system which is located above water level or exposed at low tide. These roots are specialized in their anatomy and are generally described as pneumatophores ('air bearing') or more precisely pneumorrhizae. They take a range of forms and develop in different ways, and are well endowed with lenticels (114) and internal air spaces which are continuous with those of submerged roots allowing gaseous exchange in the latter. Pneumatophores occur as prop or tabular roots (101) or develop from shallow horizontal roots as laterals that grow vertically upwards (peg roots 104). These may or may not become thickened (105a). In some cases these peg roots are themselves supported by prop roots. Alternatively the shallow horizontal root loops upward above the waterlogged level and back down again. The aerial loop or 'knee' then becomes progressively thickened, or can remain relatively thin (105b, d). The horizontal root may remain submerged, the lateral looping once only to produce the knee root (105c). A number of plants, particularly palms, growing in waterlogged conditions develop numerous very small lateral roots with a mealy appearance often on the surface of prop roots (102); these are termed pneumatorhizae; individual sites of gaseous exchange visible on the surfaces of pneumatophores are referred to as pneumathodes (Tomlinson 1990).

Fig. 104. *Rhizophora mangle*
A mangrove swamp with a tangle of prop roots (102). Pneumatophore roots of *Avicennia nitida* are developing vertically upwards out of the water in the foreground.

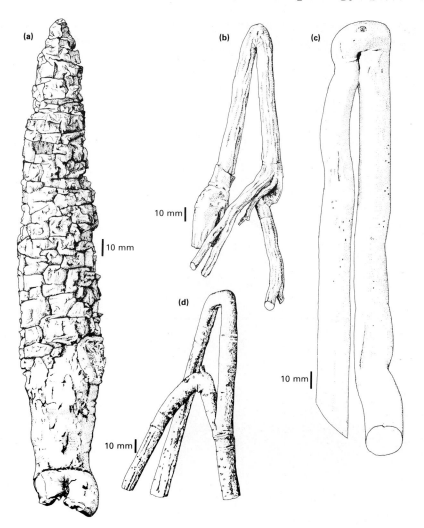

Fig. 105. a) *Sonneratia* sp.?, peg root; b) *Mitragyna ciliata*, knee root; c) *Gonystylus* sp., knee root; d) *Symphonia gabonensis*, knee root.

10 mm

10 mm

10 mm

10 mm

Fig. 106. *Myrmecodia echinata*
The swollen root tuber (110) is
chambered, and houses ants
(cf. domatia 204).

The roots of any one plant show a range of
morphologies. Some roots may be relatively thick
and tough, others very fine and fibrous. In
dicotyledons varying degrees of lignification take
place. Major roots of dicotyledonous trees can be
massive structures, possibly showing annual
growth rings in cross-section, and developing a
thick bark. Other roots show more specific
modifications. They may form prop and aerial
roots (102), breathing roots (pneumatophores
104), storage organs (tubers 110), haustoria of
parasitic plants (108), or form structures in
association with other organisms, i.e.
mycorrhizae and nodules (276). Roots can also
bear buds (178). Individual roots can shorten
considerably in length forming contractile roots
(**107e**) which maintain a corm or bulb, for
example, at a particular soil level. Contraction is
brought about either by shortening and widening
of cells or total collapse of cells. Adventitious
roots (98) of some climbing plants may branch
(**96**), expand into cavities, secrete a slow drying
cement (**99a**) which forms an attachment to the
substrate, or actually twine about a support (**98**).
Other aerial adventitious roots particularly of
epiphytic orchids are covered with layers of dead
cells, the velamen, appearing white when the
tissue is full of air. Velamen can become
saturated with water up to an inner waterproof
layer, except for small areas which remain full of
air, and then the root will appear green due to
chloroplasts in deeper tissues. However, it
appears that the aerial roots do not absorb water
from the velamen; water is absorbed only from
distal ends in contact with a substrate, the

velamen acting in a protective capacity. In a limited number of plants selected roots lose the meristematic apex and root cap and develop a woody point. Such root spines occur above or below ground in different species (**107d**).

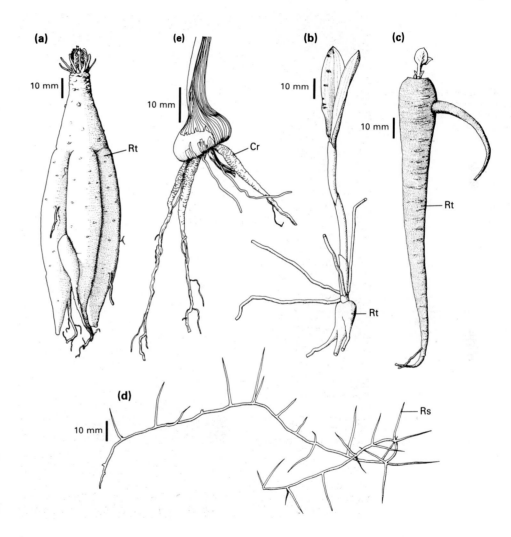

Fig. 107. a) *Incarvillea delavayi*, underground swollen roots; b) *Dactylorhiza fuchsii*, root tuber at stem base; c) *Mirabilis jalapa*, underground swollen root; d) *Dioscorea prehensilis*, branched spiny roots; e) *Crocosmia × crocosmiflora*, corm with contracted roots. Cr: contractile root. Rs: root spine. Rt: root tuber.

Fig. 108. *Cuscuta chinensis*
A swollen pad develops at intervals at points of contact of the parasite's stem with that of the host. Seen as bulges on the two lowest loops.

Parasitic and hemiparisitic flowering plants (non-photosynthetic and photosynthetic, respectively) obtain the whole or part of their nutritional requirements by the intrusion of haustoria into the host's tissue. The morphological nature of the haustoria vary considerably and cannot in most cases be unequivocally recognized externally or internally as root modifications. The extreme situation is found in *Rafflesia* in which the body of the plant consists of delicate branching threads composed of amorphous masses of cells permeating the food and water-conducting system of the host. Only the production of flowers betrays its presence. Other parasitic plants develop adjacent to their hosts; where their roots come into contact, outgrowths of the parasite attach to the surface of the host and connection is developed internally by the formation of an haustorium. The haustoria of one parasitic species can be different in structure on different host roots. Haustoria develop from the stems of climbing parasitic plants which have no contact with the ground after the initial seedling roots. Species in the Loranthaceae are typically hemiparasites, with green leaves. They form mostly woody shrubs although some species reach the proportion of small trees. Their haustoria show a number of distinctive features. The parasite may be attached at one point on the trunk or branch of a host tree (**109a, b, e**) and the host may respond by developing abnormal swellings, very elaborate ones being termed wood roses. An haustorium can consist of a single structure embedded in the host tissue (a sinker) or a number of these may develop at one point of

attachment. Alternatively, structures form which have been called epicautical roots, or runners (cf. 134), developing over the outer surface of the host (**109c**). At intervals the runner produces attachment discs, or haptera, with haustoria penetrating the host from beneath each hapteron. Runners may grow along a live branch and then turn around and return if a dead broken end is encountered. The host may die distally to the point of attachment of the parasite. Establishment (168) of seedlings of the Loranthaceae is complex. The seed is initially attached to the host branch at the hypocotyl. It is unclear whether there is any root axis present. The base of the hypocotyl swells to form a primary haustorium and is glued to the surface at this stage (in *Viscum* for example). Distortion of haustorial tissue forces a sinker into the host tissue. The seedling may now be erect with photosynthetic cotyledons. In some species the cotyledons are connate (234) and lie on the host surface. The plumule emerges from the slit between the cotyledons. Details vary considerably from one species to the next. An extensive account of the biology of parasitic plants is given by Kuijt (1969).

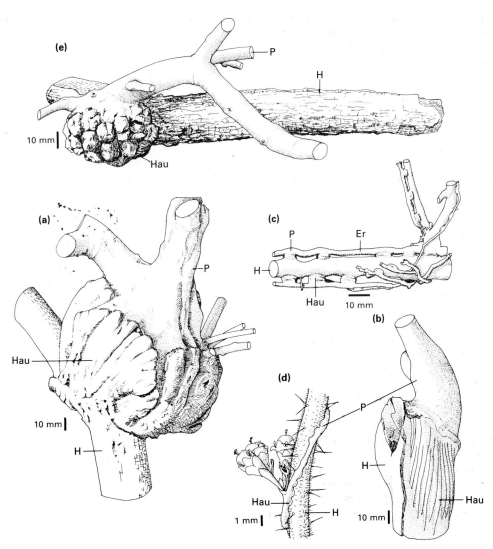

Fig. 109. Parasite/host connections. a) *Tapinanthus oleifolius*; b) *Phoradendron perrottetii* (on *Protium insigne* host); c) *Amylotheca brittenii*; d) *Cuscuta* sp. (on *Urtica pilulifera* host); e) *Lysiana exocarpi* (on *Hakea intermedia* host). Er: epicautical root. H: host. Hau: haustorium. P: parasite.

Fig. 110. *Chlorophytum comosum*
Excavated plant showing swollen root tubers. The inflorescences demonstrate false vivipary (176).

Expansion of a root laterally by cell division and enlargement gives rise in many species to a swollen root or root tuber (similar underground structures can be formed from swollen stems, 138). Frequently only a proportion of the roots on a plant will form tubers which vary considerably in different plants in their size and shape. In some orchids just one adventitious root (98) swells during each growing season providing storage material for growth after the resting period. A similar development occurs in *Ranunculus ficaria*. Here, single adventitious roots are produced at the base of buds on the aerial stem. The root swells to produce a detachable 'tubercule' which also includes the bud's apical meristem. Similar tubercules develop from adventitious buds on the stem base. In each case additional adventitious buds can develop on a tubercule itself. Thus organs are produced which are composed of tissue derived from both root and shoot (see dropper 174). In contrast to a stem tuber, a root tuber will have a root cap, at least when very young, and it may bear lateral roots but will not bear a regular sequence of scale leaves subtending buds, although there may be one or more buds present at its proximal end. These buds may be derived from the stem to which the adventitious root is attached or represent adventitious buds (232) arising from the root itself. The primary root of a plant can become swollen to produce a tap root tuber, usually in conjunction with a swelling of the base of the hypocotyl (166). Large woody swellings form on some trees and shrubs and can be partly of root tissue origin. Such woody structures are referred to as lignotubers (**138a**).

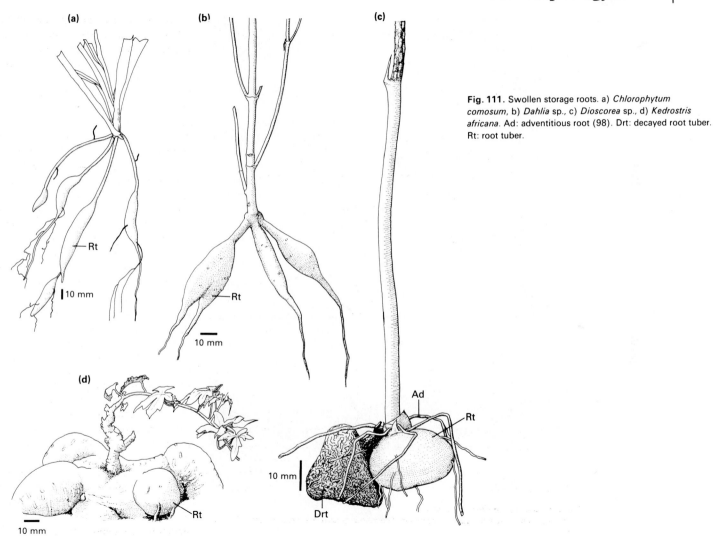

(a)

Rt

10 mm

(b)

Rt

10 mm

(c)

Ad

Rt

10 mm

Drt

(d)

Rt

10 mm

Fig. 111. Swollen storage roots. a) *Chlorophytum comosum*, b) *Dahlia* sp., c) *Dioscorea* sp., d) *Kedrostris africana*. Ad: adventitious root (98). Drt: decayed root tuber. Rt: root tuber.

Fig. 112. *Linaria* sp. Abnormal stem development. A ribbon-shaped structure (fasciation 272) instead of cylindrical.

A stem consists of a series of nodes separated by internodes. Leaves are inserted on the stem at the nodes and commonly have buds in their axils (4). (A forester uses these terms in a different fashion, a node marking the location of a whorl of branches on a trunk, the portions of trunk between whorls constituting internodes.) Internodes may be very short, in which case one node appears to merge into the next. The combined structure of stem and leaves is termed a shoot (4) and thus each bud in a leaf axil represents an additional shoot. The sequences of shoot development give any plant its particular form. Each stem grows in length owing to the activity of an apical meristem situated at its distal end. The dome of cells that forms the apical meristem is constantly changing its size and shape as new leaf primordia (18) are initiated from its flanks (exogenous development) in a regular sequence (218). Older, more proximal, leaves can form some sort of protection over younger leaves (264). The time interval between the formation of two successive leaf primordia on the apical meristem is termed a plastochron(e). The stem can increase in width just behind the apex as well as in length. This is particularly apparent in monocotyledons, especially palms, where later increase in thickness due to the activity of a cambium (16) is not usually possible. The apical meristem of a stem may produce leaf primordia continuously, or rhythmically with intervals of rest (260). Leaf production may be out of phase with stem elongation (**283i**). The apical meristem of any one shoot is sometimes referred to as the terminal meristem (or terminal bud), to distinguish it from the axillary meristems (or axillary buds) borne in the axils of its leaves. (Lateral meristem is used in a different context 16.) Each axillary meristem can develop into a shoot in its own right and will have its own apical meristem. The apical meristem may continue to function more or less indefinitely resulting in monopodial growth (250). Alternatively the apical meristem may sooner or later change its activity and terminate with the production of a flower, or whole inflorescence or other organ, or otherwise lose its meristematic capabilities (244). Continued elongation of the axis can then occur by the development of an axillary meristem usually close behind the apex. Such growth is termed sympodial (250). Stems can develop in a range of shapes (120) and surfaces can become elaborated by bark development (114), emergences (116), adventitious roots (98), and adventitious buds (232).

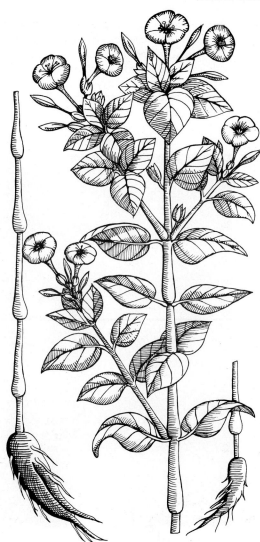

Fig. 113. The 'marvell of Peru with white floures' (*Mirabilis jalapa*) redrawn from Gerard (1633). The figure illustrates a root tuber (cf. **107c**) and stem pulvini (cf. **129**).

The surface of a stem (or root and occasionally petiole **40a**) can become elaborated by the development of a layer of bark. The bark of any one species is characteristic and is an aid to identification, although it will vary considerably depending upon the age of the trunk or stem. The term bark is often applied to the whole structure that can be pulled away from the wood. However this layer will include at its inner surface the phloem (food conducting tissue) and bast (phloem) fibres. Bark strictly applies only to the outer layer of tissue that develops from a cylinder of meristematic cells within the stem, the cork cambium (or phellogen), and which constitutes a lateral meristem (16). Cells external to the cork cambium are dead, cells internal to this cambium may contain chloroplasts—if the outer layer is thin the bark can appear green. As a stem expands in width, the dead layers of bark are forced apart to be replenished from within. In addition the cork cambium often does not form a simple cylinder in the stem but has an irregular three-dimensional arrangement such that the bark is produced in isolated sections which can become detached independently. These features give bark its variously textured appearance. The bark is punctuated at intervals by small patches of loosely packed cells allowing air to penetrate to underlying live tissues. These cell patches, lenticels, can be conspicuous at the surface particularly in smooth bark (**115a**). Bark will form characteristic patterns around the scar of a fallen branch or leaf (**115e**). The natural appearance of the bark of tropical trees is loosely described as belonging to six broad categories by Corner (1940): smooth (**115a**), fissured (**115b**), cracked (**115c**), scaly (**115d**), dippled-scaly (**115e**), and peeling (**115f**). Monocotyledons, with few exceptions, lack a lateral meristem able to produce bark but many, for example palms, develop a hard outer layer of fibres derived from old leaf veins.

Fig. 114. *Ficus religiosa*
Part of a young shoot, six internodes visible. Lenticels are conspicuous on the upper three internodes; bark formation commences at each node and is more advanced in the lower, older, internodes.

Fig. 115. Bark types. a) *Prunus maakii*, smooth;

b) *Castanea sativa*, fissured;

c) *Liquidambar styraciflua*, cracked;

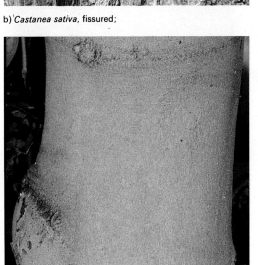

d) *Talauma hodgsonii*, scaly;

e) *Peumus boldus*, dippled-scaly;

f) *Acer griseum*, peeling.

In addition to leaves, buds, and roots, a fourth category of structure, an emergence, sometimes develops on a stem, and is usually in the form of a prickle. There is not a particularly clear distinction in the usage of the terms prickle, spine, and thorn (76). Here, prickle is used solely for a sharp structure on a leaf (76) or stem that is woody, at least when mature, and is derived from tissues just beneath the epidermis in contrast to trichomes, i.e. hairs formed from the epidermis (80). Thus, an emergence does not represent a modified stem (124), leaf (70), or root (106). Prickles occur on young stems usually in an irregular arrangement (117) and vary in size. If flattened longitudinally (117a) they may approach in appearance the winged condition of some stems (121d). On older stems the prickle may be shed leaving a scar, or persist and become a relatively massive structure (116a, 117c). Nevertheless prickles are usually relatively easily detached indicating their superficial development, and will not be expected to contain vascular tissue. They are often associated with a climbing or scrambling habit.

Fig. 116a. *Chorisia* sp.
Permanent trunk prickles.

Fig. 116b. *Aiphanes acanthophylla*
Emergences on a palm trunk; root spines (106) occur in similar locations on other species of palm (e.g. *Cryosophila* and *Mauritia* spp).

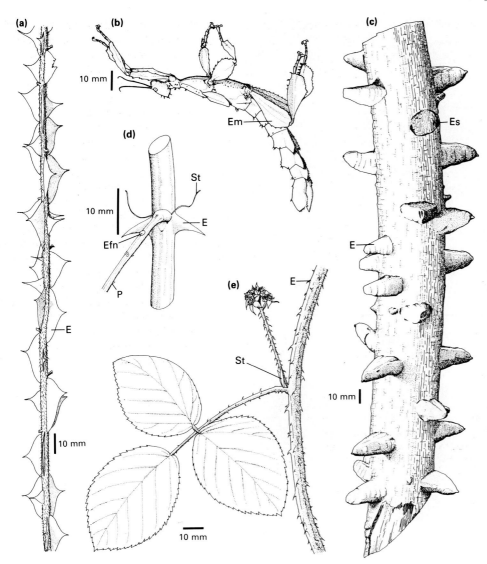

Fig. 117. a) *Rosa sericea* var. *pteracantha*, stem after leaf fall; b) *Extatosoma tiaratum*; c) *Fagara* sp., portion of old stem; d) *Acacia seyal*, stem at point of leaf attachment; e) *Rubus fruticosus* agg. E: emergence. Efn: extra-floral nectary. Em: emergence mimic. Es: emergence scar. P: petiole. St: stipule (52).

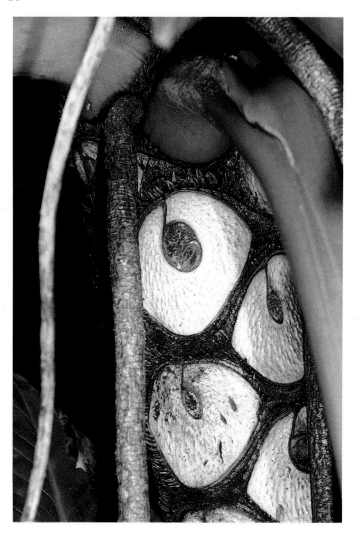

Fig. 118. *Philodendron* sp.
Each broad pale scar is that of a
detached foliage leaf. The bud that
was subtended by each leaf has also
abscissed and is represented by a bud
scar surrounded by the leaf scar.
Adventitious roots (98) also present.

Scars on stems either indicate the former position
of a structure that has fallen off, or develop in
response to injury or grafting. In young tissues
the location of injuries may be masked by
exudation of latex or resin. In old live tissue the
formation of wood and bark will produce various
structures growing over the wound. Scars left by
the abscission of leaves, roots, shoots, and fruits
will be more regular in their shape and location.
Leaves often fall due to breakage at precise points
of abscission (48) and the scar left on the stem
will indicate the former position of vascular
strands in the leaf (119a). Many plants shed
whole shoot complexes, breakage again occuring
at precise locations (268) and the corresponding
scars will remain unless subsequently enveloped
by further growth of the stem (115e). Increase in
girth will lead to the separation of scars that are
initially close together, those of a leaf and its pair
of stipules for example (119f). Stipules in many
plants abscise at an early stage in leaf
development; their existence is only detectable by
identifying the persistent stipular scars (78). The
relative position of scars on a stem can aid the
interpretation of the remaining structures (4) and
indicate for example if a shoot system is
monopodial or sympodial (250). The scale leaves
separated by very short internodes of a terminal
bud fall to leave a ring of scars indicating the
location of the bud when it was dormant. If
dormancy is a response to annual drought or
cold, the shoot system can be aged by counting
the number of rings of scars (269b).

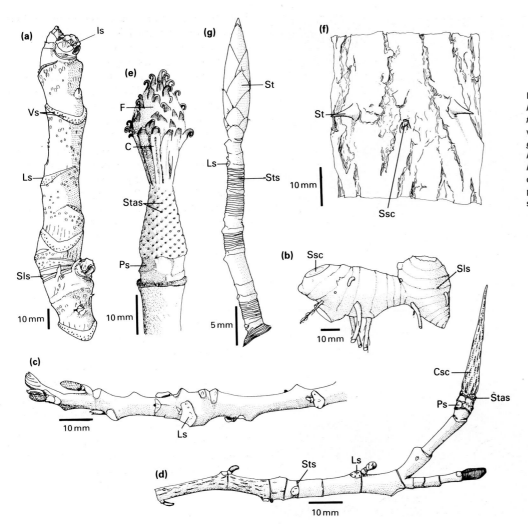

Fig. 119. a) *Aralia spinosa*, end of shoot in winter; b) *Hedychium* sp., portion of rhizome (cf. **131e**); c) *Pterocarya fraxinifola*, end of shoot in winter; d) *Liriodendron tulipifera*, winter shoot with remains of terminal flower; e) *Magnolia grandiflora*, flower after shedding of petals and stamens; f) *Robinia pseudacacia*, bark with remains of node features; g) *Fagus sylvatica*, end of shoot in winter. C: carpel. Csc: carpel scar. F: fruit. Is: inflorescence scar. Ls: leaf scar. Ps: perianth scar. Sls: scale leaf scar. Ssc: site of shed shoot. St: stipule. Stas: stamen scar. Sts: stipule scar. Vs: vein scar.

Fig. 120a, b. *Miconia alata*
Two stages in the maturation of a stem internode which is fluted. The young 'wing' tissue (a) is shed following the development of woody ridges and bark (b).

The majority of aerial stems are more or less cylindrical in shape. The herbaceous and young shoots of shrubby species of some families, typically the Labiatae, are square in cross-section becoming round if woody. Underground stems have a variety of shapes (130, 136, 138). The stems of succulent plants are typically swollen (202) and in others the stem is flattened and mimics a leaf (126). The bases of leaves in some cases are extended some distance down the stem forming ridges (24). If particularly extended the stem becomes winged or pterocaul (**121a, d, e**). In such cases leaves may fall off very soon or be represented by scales, the photosynthetic activity being confined to the green stem and its flanges. A simple cylindrical shape may become elaborated by the formation of bark (114), or in the case of climbing plants develop a range of contortions and twistings (**121c**) due to differential growth rates of different tissues and the production of areas of short-lived and easily ruptured cells. The old but living stems of some desert plants are similarly disrupted and split following the formation of longitudinal sections of cork within the wood. The trunks of some tropical trees become so deeply fluted that holes develop through from one side to the other, a condition known as fenestration.

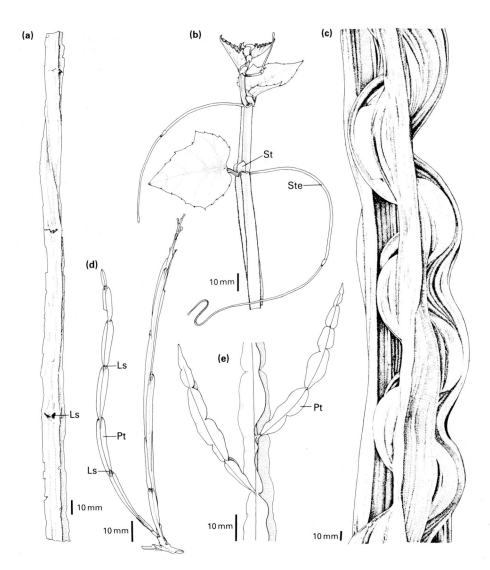

Fig. 121. a) *Cissus* sp., portion of old stem; b) *Cissus quadrangularis*; c) *Bauhinia* sp., old liane (308) stem; d) *Genista sagittalis*; e) *Baccharis crispa*. Ls: leaf scar. Pt: pterocaul stem. St: stipule (52). Ste: stem tendril (122).

Fig. 122a. *Gouania* sp.
A stem tendril bearing leaves.

Fig. 122b. *Illigera* sp.
Axillary shoots take the form of recurved hooks.

Numerous climbing plants possess tendrils, or hooks acting in the manner of grappling irons. These structures may represent modified leaves (68), parts of leaves (petiole 40, stipule 56) or be derived from stems. Prehensile stem tendrils can become secondarily thickened to form permanent woody clasping hooks (122a, 123a). Alternatively a tendril will twine around the support and subsequently shorten in length by coiling up, the proximal and distal ends of the tendril often twisting in opposite directions. Stem tendrils may be branched; some have adhesive discs at their distal ends. Stem tendrils and hooks represent either modifications of axillary shoots or are terminations of a shoot, continued growth of that axis being sympodial (250). Frequently tendrils or hooks are produced as an apparent alternative to an inflorescence and in such cases have been referred to as modified inflorescences (145b, d). Tendrils may or may not bear leaves and buds and their true identity is often difficult to interpret giving rise to different published opinions. This is particularly so for the families Vitidaceae, Passifloraceae, and Cucurbitaceae. In the latter family, the single (123e) or sometimes pair of tendrils at a node is usually taken to represent a modified leaf, a prophyll (66) although this is not confirmed for *Bryonia* by Guédès (1966). Such decisions should be arrived at after careful study of the development of the shoot at the apical meristem and in particular the location of new tendril primordia in relation to other structures, i.e. leaves and buds, being produced (4, 6). Developmental studies of the formation of the tendril in *Passiflora* species of Passifloraceae indicate that each leaf subtends a collection of accessory buds (Shah and Dave 1971; 237c). A central bud forms the tendril which is therefore a modified stem, one or more lateral buds will develop into flowers or inflorescences, and yet another bud above (distal to) the tendril may develop into a vegetative shoot. An older interpretation (Troll 1935; his Fig. 659) based on the mature morphology of members of other genera in the family is that the tendril represents an axillary shoot and the flower or inflorescence is a lateral shoot borne on it but usually without subtending leaves (145b, 238). Similar alternative interpretations are promoted to describe the tendril in the Vitidaceae. This stem tendril, frequently branched and bearing small leaves (121b), is located on the opposite side of the stem to that of the foliage leaf at the same node (121b, 123d). These plants often show a very precise sequence of nodes with and without tendrils (229b). Accounts of the *Vitis* shoot usually take the tendril to be the terminal end of a shoot and the whole axis to be sympodial (250), a precocious lateral bud extending the growth. Studies of the development of the tendril at the shoot apex indicate that it arises on the side of the apical meristem, i.e. it is not a terminal structure (Tucker and Hoefert 1968). If the axis is considered to be monopodial then three accounts are available. Either the bud that forms the tendril is initiated 180° around the stem away from the leaf that should subtend it (Shah and Dave 1970), or the tendril bud, whilst probably subtended by a leaf becomes displaced from it during stem growth and appears at the node above, a form of adnation (234) (Millington 1966; Gerrath and Posluszny 1988). Finally, the tendril is explained as an organ '*sui generis*'—a thing apart, and therefore not in need of interpretation (206)! Additional developmental studies may help; nevertheless the mature plant has a leaf opposed tendril, and the plant is always right.

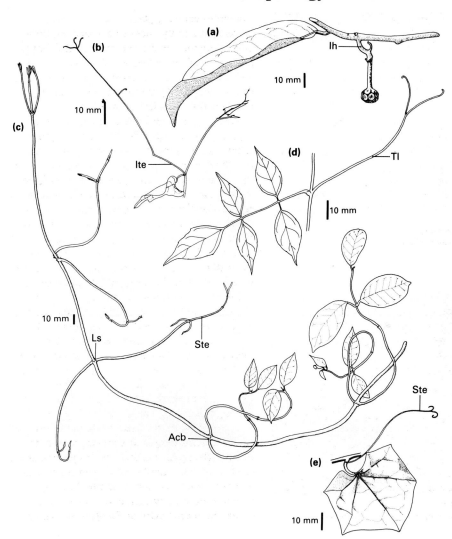

Fig. 123. a) *Artabotrys* sp., single fruit on hooked inflorescence axis (144); b) *Antigonon leptopus*, inflorescence tendril (144); c) *Hippocratea paniculata*; d) *Vitis cantoniensis*; e) *Gerrardanthus macrorhizus* (Cucurbitaceae 122). Acb: accessory bud (236). Ih: inflorescence hook. Ite: inflorescence tendril. Ls: leaf scar. Ste: stem tendril. Tl: leaf opposed stem tendril.

A spine (6) may represent a modified leaf (70), stipule (56), leaf stalk (40), root (106), or flower stalk left after the fruit has dropped (144), or may represent an emergence (76, 116), or it may represent a modified stem. There is an inconsistency in the use of the terms spine, prickle, and thorn (76). A stem spine is formed if the apical meristem of a shoot ceases to be meristematic and its cells become woody and fibrous. Such a spine may bear leaves and therefore buds which may also develop as spines (**125c, 242**), or no trace of such lateral appendages may be visible (**125a**). In the latter case the stem origin of the spine is detectable because it will be subtended by a leaf or leaf scar (6). Frequently the spine represents one of a number of accessory buds (**124a, 236b**) in the leaf axil. This is not always apparent from the mature specimen. Spines are either lateral on longer usually indeterminate (**125b**) shoots, or terminal forming a determinate shoot (**125e**). If a relatively long vegetative shoot eventually ends in a spine, only the most distal portion is referred to as a spine.

Fig. 124a. *Gleditsia triacanthos*
Two shoots (236) developing in the axil of a leaf (lost). The upper shoot is represented by a spine.

Fig. 124b. *Pachypodium lameri*
A condensed branch system (238) of spines developing in the axil of each leaf.

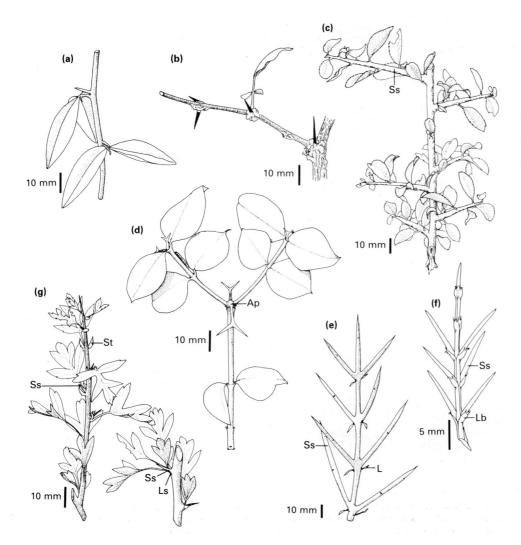

Fig. 125. Stem spines in axils of leaf or leaf scars. a) *Balanites aegyptiaca*, b) *Aegle marmelos*, c) *Prunus spinosa*, d) *Carissa bispinosa*, e) *Colletia infausta*, f) *Genista horrida*, g) *Crataegus monogyna*. Ap: apex parenchymatization (244). L: leaf. Lb: leaf base. Ls: leaf scar. Ss: stem spine. St: stipule (52).

Fig. 126a.
Muehlenbeckia
platyclados
Flattened stems of many
internodes—phylloclades
which arise in the axils of
leaves.

Fig. 126b. *Phyllanthus*
angustifolius
Flower clusters along the
edge of the flattened
stems.

The stems of some plants are flattened structures which are green and photosynthetic and bear small scale leaves. Such flattened stems are referred to as phylloclades or cladodes. A plant may be composed entirely of these structures, or they may be borne on more familiar cylindrical stems (**247a**). A phylloclade (cf. phyllode which is a flattened leaf petiole 42) consists of a stem representing a number of internodes (**126a, b, 127b**). Phylloclades can be recognized by the presence of scale leaves or scars where temporary leaves have fallen off. Buds in these leaf axils will give rise to additional phylloclades or to inflorescences (**126b**). In the case of phylloclade-bearing cacti (**127b, 203a**), the leaf/bud site is marked by an areole (202). A cladode is a flattened stem of limited growth, the apical meristem aborting, and the stem usually consists of only one or two internodes (**127a, c, d**). A cladode or a phylloclade is subtended by a leaf, which is often a scale leaf or a scar where this has dropped (**127d**). A cladode may bear a scale leaf plus subtended axillary bud on its surface (**127d′**) and may then at first sight resemble an epiphyllous leaf (**74**). A group of cladodes may appear to rise in the axil of a single scale leaf (**127a**) and then represent proliferation shoots (**239g**). Pterocauly (**121e**) describes the condition in which a cylindrical stem has extended flattened wing-like edges. The existence on the one hand of leaves bearing buds (74) and on the other hand flattened stems, usually described as phylloclades or cladodes, allows scope for considerable discussion concerning the nature of these organs. Conventional morphology will wish to fit each example into a discrete category whilst recent developmental studies advocate a continuum of expression of leaf/stem features in such organs, i.e. there is a transference of features between organ types, a phenomenon referred to as homeosis (Cooney-Sovetts and Sattler 1986; Sattler 1988).

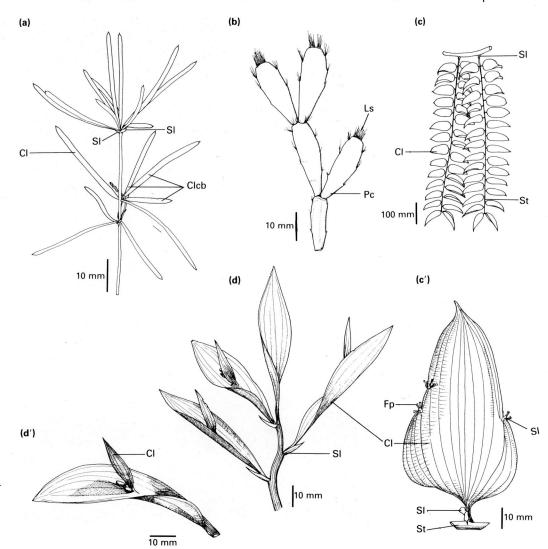

Fig. 127. Flattened green stems in the axils of scale leaves. a) *Asparagus densiflorus*, b) *Rhipsalidopsis rosea*, c, c¹) *Semele androgyna, d, d¹) Ruscus hypoglossum*. Cl: cladode. Clcb: cladode condensed branching (cf. **239g**). Fp: flower pedicel. Ls: leaf spine (202). Pc: phylloclade. Sl: scale leaf. St: stem.

Fig. 128a. *Rhoicissus rhomboidea*
An old stem pulvinus that has become enlarged and lignified (woody).

Fig. 128b. *Piper dilatatum*
Stem pulvini, swellings at every node.

A pulvinus is a swollen joint on a stem or leaf. In the latter case a distinction can be made (46) between a pulvinus which allows reversible changes in orientation, a pulvinoid which allows irreversible movement, and an articulation joint marking a point of future breakage. Articulation joints occur on stems resulting in stem shedding (268), identified by the presence of scars plus fallen stems, but swollen stem joints allowing movement are mostly of the pulvinoid type, i.e. bending at the joint is likely to be due to cell division in a meristematic region (112) and therefore to be non-adjustable. Nevertheless many stems bend at a pulvinus if wilting and then recover the original position if watered. In these cases rigidity is maintained by turgidity, mechanical tissue being largely absent whilst the joint remains meristematic. It is not clear if any species possess a pulvinus that does allow repeated bending one way and the other in the manner of a leaf pulvinus. A stem pulvinus can become considerably enlarged with growth (**128a**) and eventually become woody (lignified).

Fig. 129. *Mirabilis jalapa*. A pulvinus occurs at the proximal end of most internodes.

Fig. 130. *Alpinia speciosa*
An excavated rhizome system. The underground portion of each sympodial unit (250) persists considerably longer than the distal aerial portion which is shed at an abscission zone (268). Model of Tomlinson (295d).

A stem growing more or less horizontally below ground level is described as a rhizome (170). Rhizomes tend to be thick, fleshy or woody, and bear scale leaves or less often foliage leaves (87c), or the scars when these leaves have been lost; they also bear adventitious roots most frequently at the nodes. Rhizome diameters vary from a few millimetres in some grasses up to half a metre or more as in the palm *Nypa*. A root bearing root buds (178) can be distinguished from a rhizome by the lack of subtending leaves or leaf scars. The majority of rhizomes are sympodial (250) in that the distal end of each shoot becomes erect and grows vertically bearing foliage leaves and usually terminal or lateral flowers, although in some plants an inflorescence develops directly from a bud on the underground rhizome. Growth of the rhizome is continued underground by the activity of one or more axillary buds the development of which may be seasonal and by which the rhizome is therefore ageable. The relative position of successive sympodial units (250) can be quite regular (269d). The aerial portion of each sympodial unit is usually only temporary and abscises at ground level (unless it forms a climbing axis), the rhizome portion persisting for some time before eventually also rotting (171c, d, 130). Monopodial rhizomes, having lateral aerial shoots, probably die at their distal end eventually, and again continued rhizome extension must then result from an axillary bud. A sympodial rhizome may appear superficially to be monopodial due to adnation (235a). Rhizomes in which each sympodial unit is relatively short and fat are described as pachycaul (**131b**), long and relatively thin rhizomes conversely are described as leptocaul (**131f**, cf. **194**). (These terms are also applied to aerial plant stems.) A particular type of rhizome, growing out from the base of an otherwise erect plant and consisting of a single underground horizontal stem turning erect at its distal end is a sobole. Examples of soboliferous plants occur in the Palmae and the Araceae. Although rhizomes are typically defined as being horizontal shoot systems, there are many examples where the shoot system is in fact vertical growing either upwards or downwards and may indeed develop from above-ground parts of the plant.

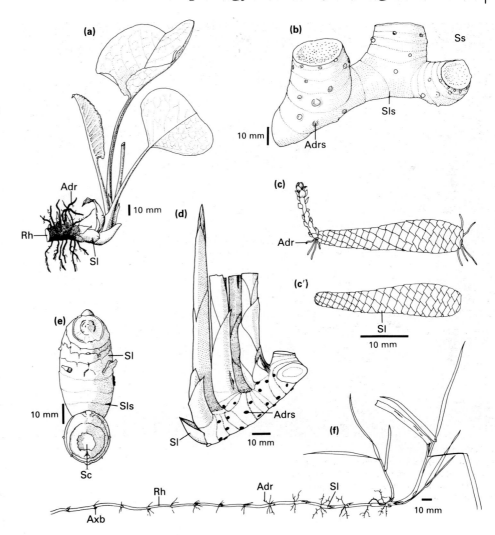

Fig. 131. a) *Petasites hybridus*, rhizome bearing foliage leaves; b) *Costus spiralis*; c, c¹) *Achimenes* sp.; d) *Cyperus alternifolius*, young end of rhizome (cf. **269d**); e) *Hedychium* sp., top view rhizome (cf. **119b**); f) *Agropyron (Elymus) repens*. Adr: adventitious root. Adrs: adventitious root scar. Axb: axillary bud. Rh: rhizome. Sl: scale leaf. Sls: scale leaf scar. Ss: stem scar.

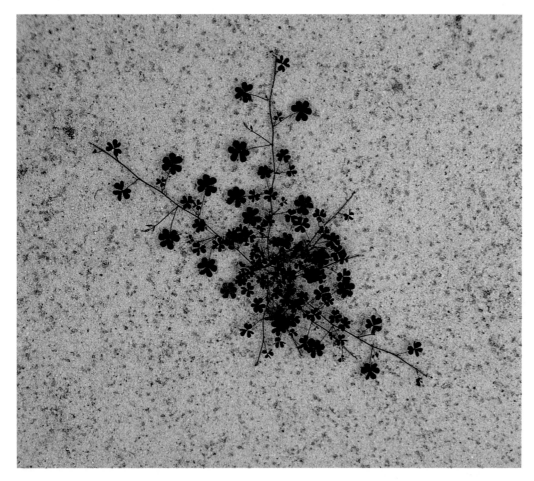

A loose distinction exists between the definition of a stolon and a runner (170, 134). A stolon is a stem growing along the substrate surface or through surface debris. It has long thin internodes and bears foliage, or occasionally scale leaves. Buds in the axils of the leaves will develop into inflorescences or additional stolons. Adventitious (98) roots usually emerge at the nodes (nodal roots), sometimes only at nodes having a lateral stolon. Stolons develop in radial fashion from a young seedling (132) and then fan out, rooted nodes forming young plants as and when connecting stolons become damaged or decayed (171a, b). Stolon growth may be monopodial or sympodial (250). In *Echinodorus* (Charlton 1968) the main upright axis of the plant is sympodial, the evicted end of each sympodial unit becoming a horizontal stolon that continues to grow monopodially although in certain conditions an inflorescence is produced instead of a stolon. The sequence of leaf and bud production in this plant is very regular.

Fig. 132. *Oxalis corniculata*
Plan view of seedling plant. The seedling axis bears leaves in eight vertical rows (221b) and their axillary buds develop into stolons diverging in up to eight potential directions.

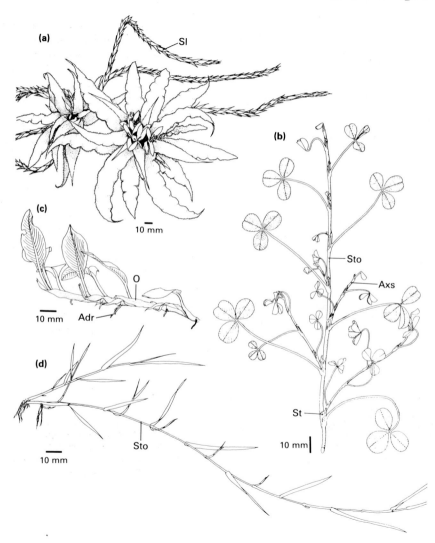

(a)

Sl

(b)

Sto

Axs

St

10 mm

(c)

O

10 mm

Adr

(d)

Sto

10 mm

10 mm

Fig. 133. a) *Cryptanthus* 'cascade', leaf rosettes with axillary shoots emerging as stolons; b) *Trifolium repens*; c) *Polygonum affine*, young end of stolon; d) *Agrostis stolonifera*. Adr: adventitious root. Axs: axillary shoot. O: ochrea (54). Sl: scale leaf. St: stipule (52). Sto: stolon.

A runner is a thin horizontal stem above ground consisting of one or more long internodes at the distal end of which is a rosette of foliage leaves or a heteroblastic (28) sequence of leaves and from which additional runners diverge. A runner does not root at any node present between the mother plant and the daughter plant; any leaves present on the runner will usually be scale leaves. The runner is often short-lived and the production of runner and rosette represents a system of vegetative multiplication (170). In a number of plants, erect inflorescences bearing bulbils (172) or bulbs with adventitious roots instead of flowers (**173d**) are present, and these stems arch over depositing potential new plants on the substrate (**177a**). These structures are equivalent to runners with the exception that the latter grow plagiotropically (246) from their inception. Similar structures, droppers (174), occur in a number of bulb (84) forming species but these descend into the ground and do not develop along the surface.

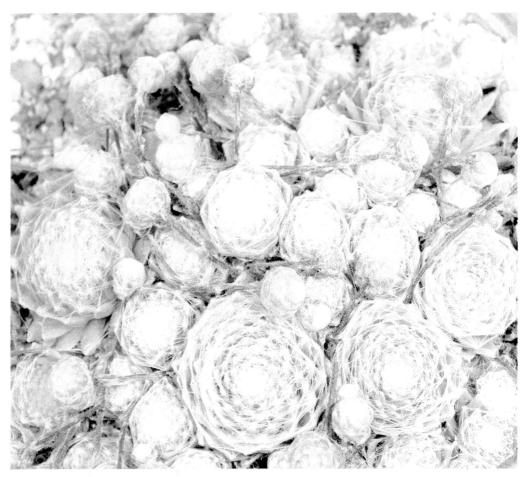

Fig. 134. *Sempervivum arachnoideum*
A new daughter rosette of succulent (82) and hairy leaves is produced at the end of each runner.

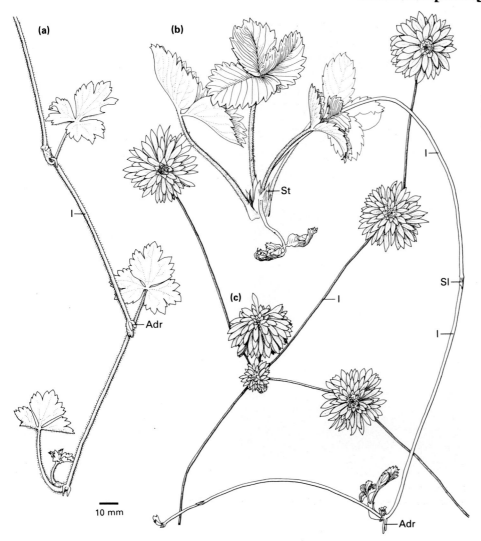

Fig. 135. Runners formed from long internodes. a) *Ranunculus repens*, b) *Fragaria × ananassa*, c) *Androsace sempervivoides*. Adr: adventitious root. I: internode. Sl: scale leaf. St: stipule (52).

10 mm

Fig. 136. *Cyanastrum hostifolium*
A dormant corm developed from a bud on the top of last season's corm. The concentric rings are the scars (118) of detached leaves. Each scar has one bud in its axil.

A corm is a short swollen stem of several internodes and nodes bearing either scale or foliage leaves. It develops at or below ground level in a vertical position. In favourable growing conditions the apical meristem (16) of the corm or one of the buds close to the apex extends into an aerial flowering shoot usually bearing foliage leaves (137b). The corm may be reduced in size at the expense of this shoot or may shrivel away altogether. One or more buds in the axils of leaves on the corm swell to form new corms during the growing season. These buds may be distally situated at the top of the old corm (171e, f), or may be proximal and laterally placed near the base of the old corm (171g, h). If the old corm persists, a vertical sympodial system of corms results (137c′). There is very little difference between this structure and a rhizome system with very short rhizome sympodial units (181d, f). Adventitious roots develop usually from the base of the corm only. These may be contractile (107e). In a few plants forming corms, the process is not sympodial but monopodial, the lowest internodes at the base of the flowering stalk swelling during the growing season to produce a corm directly on top of the previous corm, e.g. *Oxalis floribunda* (Jeannoda-Robinson 1977). The pseudobulb of an orchid (137d, 199d, f) which consists of one or several swollen internodes is equivalent to a corm.

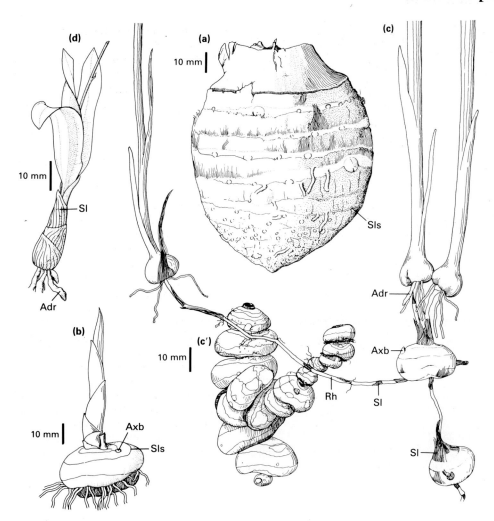

Fig. 137. a) *Colocasia esculenta*, dormant corm; b) *Gladiolus* sp., corm with emerging shoot; c, c¹) *Crocosmia × crocosmiflora*, corm sequences with and without intervening rhizome portions; d) *Polystachya pubescens*, orchid pseudobulb (198). Adr: adventitious root. Axb: axillary bud. Rh: rhizome. Sl: scale leaf. Sls: scale leaf scar.

A stem tuber is a swollen shoot usually underground and bearing scale leaves, each subtending one or more buds which give rise to vegetative shoots. The presence of leaves or leaf scars distinguishes a stem tuber from a root tuber (110). Typically a stem tuber forms by the swelling of the distal end of a slender underground rhizome and thus does not form one unit of a sympodial sequence as is commonly found in a rhizome (130). However, the distinction between stem tuber and rhizome is not always easy to apply (**139b**). A tuber of *Solanum tuberosum* (**139e**), for example, will produce a series of additional tubers if forced to develop in the absence of water (**271g′**). Stem tubers normally survive longer than the main plant and sprout at axillary buds in favourable conditions. They will also bear adventitious roots (98). Stem tubers occur on the aerial shoots of some plants (typically climbing or trailing plants) and are easily detached, then producing adventitious roots and new vegetative growth and may represent either detachable axillary buds (**139a**) or swollen nodes as in *Ceropegia woodii* (**83d**) or a pair of internodes as in *Vitis gongylodes* and *Cissus tuberosa* (**139c, 138b**). A number of woody plants, which may or may not have the potential to reach tree proportions, form relatively large stem swellings at or below ground level. These are particularly typical features of many species of *Eucalyptus* and are called lignotubers (**138a**). A lignotuber incorporates many clusters (fascicles) of dormant buds (**237d**) embedded in the bark. These bunches of buds develop into groups of new shoots after adverse

Fig. 138a. *Eucalyptus* sp.
A persistent woody stem tuber (lignotuber) bearing numerous fascicles of dormant buds. Photograph courtesy of J. C. Noble.

Fig. 138b. *Cissus tuberosa*
Internodes swollen to form a stem tuber.

conditions, such as fire in the case of *Eucalyptus*. The production of new shoots from old woody tissue in general is referred to as epicormic branching, cf. cauliflory (240).

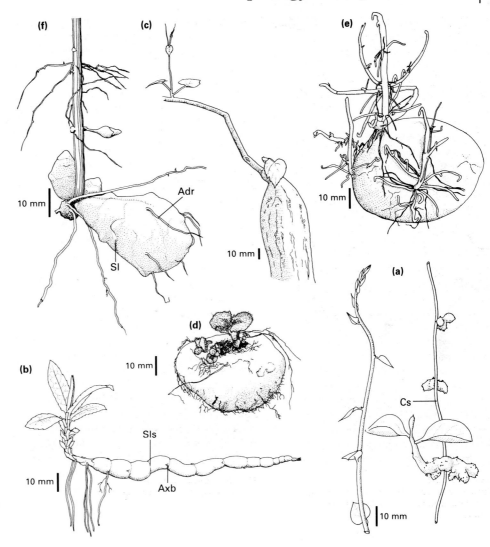

Fig. 139. a) *Anredera gracilis*, tubers on aerial climbing shoot; b) *Ballota nigra*, underground swollen stem; c) *Cissus tuberosa*, part of aerial swollen stem (cf. **138b**); d) *Sinningia speciosa*, sprouting stem tuber; e) *Solanum tuberosum*, sprouting stem tuber; f) *Helianthus tuberosus*, underground swollen stem. Adr: adventitious root. Axb: axillary bud. Cs: climbing shoot. Sl: scale leaf. Sls: scale leaf scar.

An inflorescence is a reproductive shoot system bearing flowers (infructescence, once fruits are set). The inflorescence is composed of a system of branches the principal axis being termed the peduncle, or r(h)achis particularly in grasses 184; each new branch or ultimately flower arises in the axil of a leaf (a bract 62) which is frequently dissimilar to foliage leaves on the same plant. Some or all of these bracts may be absent. In some plants the inflorescences are distinct and are readily distinguished from vegetative growth. Conversely the vegetative part of the plant may merge imperceptibly into a reproductive part and the boundaries of an inflorescence as such be difficult to delimit. In some cases the whole plant may be described as an inflorescence (preface). A great deal has been written about the typology (study of types) of inflorescences (142) and a number of standard arrangements of flowers within inflorescences are commonly recognized. However, many combinations and intergradations between types exist. Inflorescences are essentially three-dimensional and often show considerable symmetry (142) and this again allows almost limitless varieties of construction. In general, a vegetative axis bearing lateral flowers is said to be pleonanthic; one terminated by a flower or inflorescence is said to be hapaxanthic. These positional differences will affect the branching construction of the plant. Thus a main distinction is usually made between branching systems that are predominantly sympodial (250) in their development—cymes, although these may incorporate monopodial components, and racemes

which are largely monopodial having one (or more) main axis in the inflorescence framework bearing a number of lateral branches or flowers. These basic permutations are further elaborated by the possible presence of more than one branch at a node, the sequence in which flowers mature (142), and the presence or absence of pedicels (flower stalks). There may be a consistent pattern in the relative length of branches and the concept of acrotony and basitony (248) can be helpful.

The most commonly used descriptive terms are as follows (Fig. 141). *Racemes* have one central monopodial axis with (**141a**) or without (**141b**) a terminal flower and bearing lateral flowers or small bunches of flowers. For flowers that are without pedicels use *spike*, a monopodial axis bearing sessile flowers (**141c**) or condensed flower clusters (**141d**). If the axis is distinctly fleshy use *spadix* (**141e**), the whole structure often being subtended by a large bract, a spathe (**74**). If the spike hangs down under its own weight it is termed a *catkin* (**141f**) and its flowers are usually of one sex. If the lateral branches of a raceme are themselves branched, then the inflorescence is termed a *panicle* (**141g**) especially if the branching is not compact. Inflorescences in which all the flowers are displayed more or less at one horizontal level are described as either a *corymb* (**141h, i**) in which the flower stalks do not originate at one point, or an *umbel* (**141l**) in which all the flower stalks arise at or near one point. The most distal components of an umbel may themselves be branched (**141m, n**). In a *capitulum* (**141j**) the flowers sit on a flattened top of the inflorescence axis which is then termed the

receptacle (cf. floral receptacle 146) and may be folded inwards to form a hypanthodium (**141k**). Such an inflorescence might look like a single flower (a *pseudanthium* **151f**). A *cyme* is constructed sympodially. In its simplest form it consists of a series of flowers each borne in the axil of the bracteole (62) of a preceding flower (**141s–v**). Such a cyme is termed *monochasial*. If each unit (article 286) of the sympodium bears two flowers, the cyme is *dichasial* (**141o**), more than two, *pleiochasial* (**141p**). If such sympodial sequences are arranged along a single monopodial axis the inflorescence is termed a *thyrse* (**141q**), or a *verticillaster* (**141r, 142**) if lateral branches occur in whorls. The three-dimensional arrangement of sympodial units in even a simple monochasial cyme is variable and difficult to describe. It is conventional to illustrate the four main variants by means of a side view which tends to confuse the three-dimensional subtleties (**141s, t, u, v**) coupled with a plan view in the manner of a floral diagram (**9d, e, f, g**). These four types are referred to as a *rhipidium* (**141s, 9d**), a *drepanium* (**141t, 9e**), a *cincinnus* (**141u, 9f**), and a *bostryx* (**141v, 9g**). Descriptions of the complexities and symmetries of more elaborate inflorescences are really only possible by means of clear three-dimensional diagrams designed to illustrate a particular aspect under consideration, such as sequence of flower opening, movement of pollen, access by insect, and dispersal of seeds and fruits.

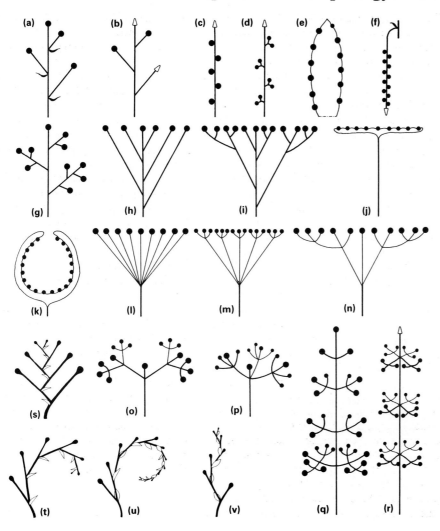

Fig. 141. Diagrammatic representation of inflorescence types. a, b) raceme, c, d) spike, e) spadix, f) catkin, g) panicle, h, i) corymb, j) capitulum, k) hypanthodium, l–n) umbel, o) dichasial cyme, p) pleiochasial cyme, q) thryse, r) verticillaster, s–v) monochasial cymes (s, rhipidium; t, drepanium; u, cincinnus; v, bostryx).

Fig. 142. *Alisma plantago-aquatica*
View oblique from below. A highly organized three-dimensional inflorescence in which branching details are repeated at increasing levels of complexity, i.e. paracladia. Model of Massart (**291i**).

Many inflorescences (140) are elaborately but uniformly branched. The conventional descriptive categories do not convey the subtleties of branching within an inflorescence or between inflorescences in a plant. Nevertheless the morphological architecture of inflorescences repays attention and is as important to the biology of a plant as is the architecture of its vegetative framework (280). A concept useful in deciphering the branching of an inflorescence is that of paracladia. A paracladial relationship ('subsidiary and similar branching') refers to the occurrence of a regular branching sequence which is repeated to build up a complete branched structure. In the developmental branching sequence shown in Fig. **143a–d**, each daughter branch repeats the behaviour of the original parent branch. The lower lateral branch system in Fig. **143c** has the same architecture as that of the whole parent branch at the previous stage (**143b**). Each such repeated unit of the pattern is termed a paracladium. The paracladial components of a typical panicle (**143g**) are shown in Fig. **143h**, outlined by dotted lines. Each paracladium need not be a complete or exact copy of the parent shoot. The paracladial construction of a dichasial cyme (**141o**) is shown in Fig. **143e**. It follows that each paracladium is itself constructed of lower order paracladia. The recognition of the paracladial relationships in the reproductive part of a plant (Troll 1964; Weberling 1965) has led to clear distinctions between two basic inflorescence types. Firstly inflorescences in which the lateral axes all terminate in a flower as eventually does the main axis, are described as monotelic inflorescences ('one direction' **143f**). All cymes (140) being sympodial are monotelic, as are racemes (140) that have a terminal flower, each repeated unit with a terminal flower representing one paracladium (**143e**). Secondly, polytelic inflorescences are described as those in which the major axes fail to terminate in flowers (**143h**). The flower bearing parts of paracladia of polytelic inflorescences are given specific names indicating the progressive development of the whole reproductive part of the plant (**143g**). Thus, that of the main axis is termed the florescence, a paracladium repeating the pattern of the florescence is a co-florescence, and incomplete florescences are termed partial florescences. The system as a whole is referred to as a polytelic synflorescence.

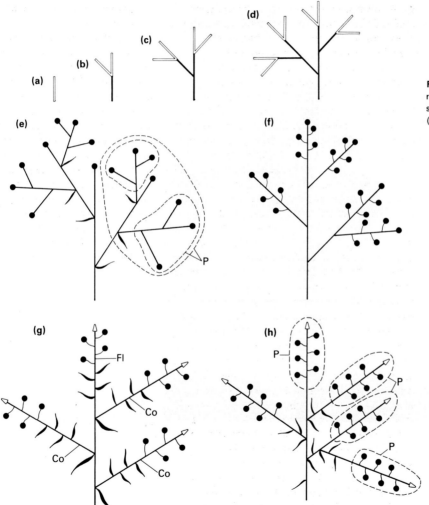

Fig. 143.)a–d) paracladial development, e) paracladial relationships, f) monotelic synflorescence, g, h) polytelic synflorescence. co: coflorescence. Fl: florescence (Co + Fl = synflorescence). P: paracladium.

Fig. 144. *Mutisia retusa*
The inflorescence, a capitulum (**141j**) subtended by an involucre of bracts (**62**) is borne on an axis which twines around supports. This plant also has leaf tip tendrils (**68b**).

An inflorescence is a branch complex displaying flowers (140). However, inflorescences may show various degrees of modification relating to alternative functions. Three general groups can be recognized. Firstly, the inflorescence bears flowers but also performs an additional activity. The inflorescence of *Bowiea volubilis* (**145c**) is a climbing organ twining around supports and has repeated wide angle divaricate (**257d**) branching each branch being subtended by a minute scale leaf. The inflorescence stem is green and replaces the photosynthetic function of leaves which are very short-lived (**85c**). The distal ends of most branches become parenchymatized (244) and cease to grow, only the last formed branches terminating in solitary flowers. Inflorescences of some other climbing plants act as tendrils (**145b, 144**) or bear hooks similar to those of other modified stems (**123a**). Secondly, an inflorescence may perform an additional function after the flowering sequence has passed. For example, the pedicels of fruits may be persistent and woody operating as permanent spines (*Euphorbia*), or, as in *Montanoa schottii*, the pedicel becomes curved and rigid forming a climbing hook once the fruit is formed. In *Bougainvillea* the pedicels of some flowers become woody and again act as hooks (**145d**). *Bougainvillea* also illustrates the third category, in which an axillary meristem may develop either into an inflorescence or into a spine (**236b**). In some climbing palms (e.g. *Calamus* species) an inflorescence may be replaced by a slender stem bearing hooks. This structure is termed a flagellum and is adnate (234) to the internode above the leaf in the axil of which it actually arises. Inflorescence penduncles or flower pedicels may perform additional functions without becoming modified. Pedicels of *Arachis* (ground nut) and *Cymbalaria muralis* elongate and bend depositing the fruit underground or into dark corners (**267c, c¹**). Similarly the peduncle of *Eichhornia* bends, placing the infructescence (**267a**) under water. The inflorescences of a number of plants bear rooting buds or bulbils in addition to or in place of flowers (**176**). Such inflorescence disruption can occur as a teratology (**145e, e¹, 270**). Inflorescences may resemble single flowers (generally termed a pseudanthium). The inflorescence paracladia (**142**) of *Euphorbia* typically consist of a single female flower lacking perianth segments and five cymes of male flowers each consisting of one stamen only (**151f**). What appear to be sepals or petals are 5 bracts alternating with 4 (or 5) pairs of fused stipules. Each such structure, a cyathium, resembles a single flower and several may be aggregated together into a symmetrical group (**8**).

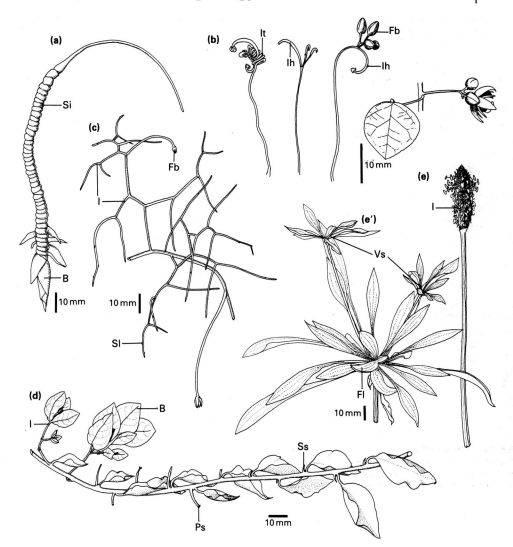

Fig. 145. a) *Coelogyne* sp., inflorescence axis with swollen internodes; b) *Ophiocaulon cissamepeloids*, variation in inflorescence modification on one plant (Passifloraceae 122); c) *Bowiea volubilis*, divaricate (**257d**) climbing inflorescence; d) *Bougainvillea* sp., climbing stem; e, e¹) *Plantago lanceolata*, normal and abnormal inflorescence (**270**). B: bract. Fb: flower bud. Fl: foliage leaf (in position of bract). I: inflorescence. Ih: inflorescence hook. It: inflorescence tendril. Ps: peduncle spine. Sl: scale leaf. Si: swollen inflorescence internode. Ss: stem spine. Vs: vegetative shoot (in place of flower).

The details of floral morphology form the basis of flowering plant classification and are therefore well documented in botanical textbooks and floras. Flowers of different species show considerable variation in morphology and only the most basic features are dealt with here. Terminology, particularly relating to the ovary, is inconsistent; that of Davis and Cullen (1979) is adopted here. The various components of a flower may be considered to be attached in sequence along a usually very short and variously shaped central axis, the torus or floral receptacle (**147a**) (cf. inflorescence receptacle 140). Due to the shortness of the receptacle the most distal female organs, collectively called the gynoecium, appear to be in the centre of the flower and to be surrounded by the more proximal male organs (androecium) which are in turn surrounded by perianth segments (such as petals and sepals). If the axis between the gynoecium and the androecium is elongated it is termed a gynophore (**147b**). In a male-only flower the gynoecium is absent or represented by a non-functional pistillode; in a female only flower the androecium is missing or present as non-functional staminodes. The sequence of gynoecium/androecium/perianth is not always apparent, (**147j–o**).

The perianth segments represent a variable number of variously coloured or green modified leaves arranged spirally or in whorls. Occasionally the perianth segments are separated from the sexual organs by an elongated portion of axis termed an androgynophore (**147c**). The perianth segments are often variously joined with each other (connation 234) or with parts of the androecium or gynoecium (adnation 234). If whorls of distinctly different perianth segments are present, the inner more distal components, the petals, constitute the corolla, and the outer more proximal components, the sepals, represent the calyx (**147d**). If all the perianth segments are similar in appearance they are individually termed tepals. Additional leaf-like structures may be found proximal to the calyx. These may take the form of whorls of bracteoles, an involucre (**147e**) (also applied to bracts of some inflorescences). A single whorl of bracteoles inserted closely below the calyx is termed an epicalyx (**147f**); in some cases the epicalyx represents the stipules of the adjacent sepals. The gynoecium consists of one or more carpels borne on the receptacle. In a relatively small number of plant families the many carpels are not joined together (free) and the gynoecium is said to be apocarpous (**147g**). Each such carpel contains a cavity, the ovary (or 'cell') into which develop one or more ovules each borne on a stalk (funicle). The funicles arise in specific regions termed placenta and each ovule will become a seed if fertilized. The term ovary is nowadays used to indicate both cavity and carpel walls which will form all or part of the fruit (154) when the seeds develop. Attached to each carpel is a style, a structure supporting a surface receptive to pollen, the stigma. Each carpel, style, and stigma of an apocarpous flower is collectively sometimes called a pistil (**147g**). In the majority of plants the carpels, frequently 3, 4, or 5 in number, are united together and the gynoecium is termed syncarpous. The styles of the carpels may remain distinct or be more or less united into one structure (**147h**). Transverse and longitudinal sections cut through various syncarpous gynoecia will show a range of distinctive configurations of carpels and different locations of placenta. The composite structure is referred to as the ovary although again this term was originally applied to the ovule containing cavities only. The whole syncarpous gynoecium is termed a pistil. Each carpel may retain its own cavity (**147i**), or loculus, or the cavities may variously merge into one another (**147i^1**). The syncarpous ovary is recorded as multilocular, trilocular, unilocular, etc. according to the number of cavities present. These aspects of gynoecium construction are included in floral diagrams and formulae (150).

The androecium consists of a number of stamens. Each stamen is composed of an anther containing pollen and is supported on a stalk, the filament (**147c**). Stamens may be united (connate 234) into one or more tubes or groups (mon-, di-, polyadelphous) or variously attached (adnate 234) to petals or gynoecium. A number of distinctive juxtapositions of ovary, stamens, and perianth segments are recognized. If the ovary is sited above the stamens, and the stamens are attached above the perianth, the ovary is described as superior (**147a–h**). Alternatively, the ovary may be over-topped by the edges of the receptacle on which it sits such that the stamens and perianth segments are positioned above the ovary. The ovary is then inferior and the stamens and perianth segments are epigynous (**147j**). If the epigynous calyx, corolla, and stamens (**147k**), or corolla and stamens (**147l**) are united together then the common supporting structure is referred to as an epigynous zone. In a flower with a superior ovary, the stamens and perianth segments are hypogynous (**147h**). However, stamens, corolla, and calyx associated with a superior ovary may be variously united and then supported on a perigynous zone; peri ('around') because this structure usually holds the stamens or perianth alongside the ovary (**147m**). In Fig. **147n** the calyx is hypogynous whilst the corolla and androecium are borne on a perigynous zone. If the stamens and perianth are actually attached to the side of the ovary (partly epigynous) the flower is described as partially inferior (**147o**). The manner in which petals and sepals overlap with one another (aestivation 148) is characteristic and is recorded in the floral diagram (150).

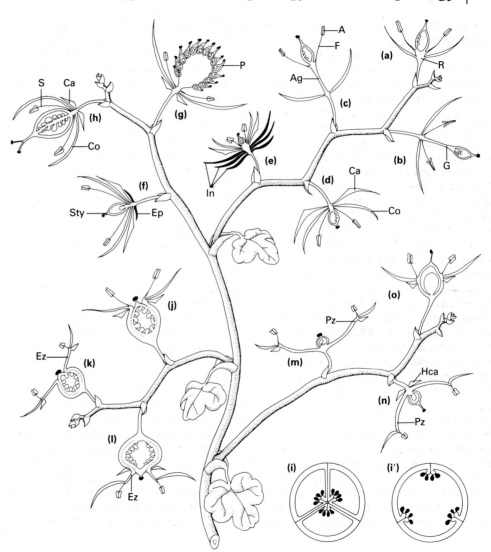

Fig. 147. Diagrammatic representation of flower structure. a)–h) superior ovary (g, apocarpous, h, syncarpous); i) trilocular ovary (transverse section); i¹) unilocular ovary (transverse section); j)–n) inferior ovary; o) partly inferior ovary. A: anther. Ag: androgynophore. Ca: calyx. Co: corolla. Ep: epicalyx. Ez: epigynous zone. F: filament. G: gynophore. Hca: hypogynous calyx. In: involucre. P: pistil. Pz: perigynous zone. R: receptacle. S: stamen. Sty: style.

Fig. 148. *Malvaviscus arborea*
The petals show convolute aestivation (cf. Figs. **39e** and **151d**). The upper flower has its petals spiralling in the opposite direction to the two lower flowers. There is a twisted staminal tube surrounding the style, the latter protruding above as a much branched stigma (146).

The various organs and perianth segments of a flower are usually arranged in a predictable manner for any given species and show various degrees of symmetry and of packing of parts in the bud (aestivation). The same terminology is employed as for the folding (ptyxis 36) and packing (vernation 38) of leaves in a vegetative bud. These aspects are usually described by means of a stylized floral diagram (150) in which certain conventions of orientation are observed. For lateral flowers, the location of the supporting stem axis is indicated together with the position of the subtending bract (**149a**). The side of the flower nearest the axis is referred to as posterior (or upper), that nearest the bract as anterior (or lower). Planes of symmetry through the flower, as seen from above (**149b**) are described as vertical, median (or transverse), and diagonal; any other plane is oblique. If an imaginary slice down through the flower results in two similar sides then the flower is symmetrical (otherwise asymmetrical **151d**). Symmetrical flowers may have only one plane (usually vertical but occasionally median or diagonal) in which they may be cut to give matching sides. Such flowers are zygomorphic (**151c, e**). If two or more cuts can be made down through the flower to give similar sides, then the flower is actinomorphic (**151a**). The matching pairs produced by different planes of slice are not necessarily the same. Bracts and bracteoles are not taken into account. The arrangement and packing of perianth segments in the flower (aestivation or prefloration) is best seen in a horizontal section cut through the flower bud. If the proximal ends

of the segments are attached to one another (united) the aestivation can apply to the unattached (free distal ends) only. Terminology follows that for vegetative leaves in a bud (vernation 38), and applies to the perianth segments in one whorl: open (**39c**), e.g. sepals not meeting at their edges; valvate, e.g. sepals touching at their edges (**39b**); crumpled (**149c**); imbricate, e.g. sepals with overlapping edges (**149d–j**). Different types of imbricate aestivation are convolute (or contorted) with segment edges alternately tucked in or out (**39e, 148**). When five perianth segments occur in one whorl the term quincuncial is applied if two are outside, two are inside, and one is half in, half out. A number of permutations are possible depending upon which segment edges overlap which (**149e, f, g**). Other arrangements of five segments exist which do not fit the quincuncial or convolute conditions (**149d, h, i, j**). If each segment in a whorl overlaps the one posterior to it, the aestivation is ascending (**149i**), or descending (**149j**) if the reverse is true. The symmetry of the flower as a whole and the relationship of the segments of one whorl (e.g. petals) to another (e.g. androecium) is recorded in the full floral diagram (150). There are extensive published accounts of the nuances of aestivation, for example Schoute (1935).

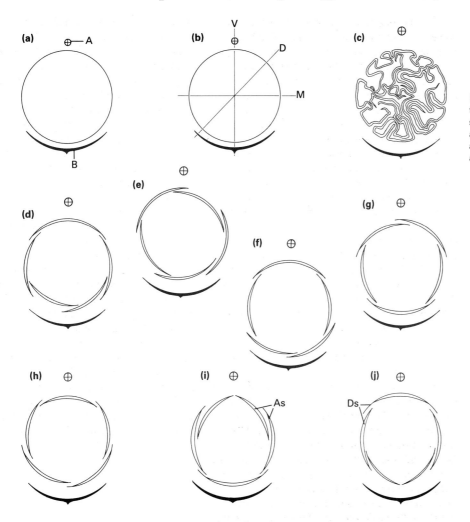

Fig. 149. Diagrammatic aspects of aestivation. a) location of flower (circle) in relation to axis and bract; b) planes of symmetry; c) crumpled aestivation; d)–j) imbricate aestivation (e^1, f^1, g: quincuncial). A: axis. As: ascending aestivation. B: . bract. D: diagonal. Ds: descending aestivation. M: median. V: vertical.

The consistent constuction and degree of symmetry (148) of flowers allows them to be conveniently portrayed by means of conventional stylized floral diagrams (8) and summarized in a floral formula. The diagram views the flower from above and shows its orientation with respect to the stem to which it is attached and the bract in whose axil it sits (149a). The location of bracteoles if present is also indicated (151a). The various parts of the flower which originate on the flower axis in the ascending order calyx, corolla, stamens, ovary (146) are portrayed in one horizontal plane with the calyx to the outside and the ovary in the centre. If the perianth segments (146) can be clearly differentiated into sepals and petals this difference is indicated by using different shaped symbols (151a). Each individual stamen (collectively the androecium) is represented by a symbol indicating the side on which the anther opens to release pollen, outwards (extrorse 151a), or inwards (introrse 151b). The gynoecium is shown in the form of a transverse section through the ovary to illustrate the carpel arrangement. If various parts are attached to one another, this is indicated by appropriately placed lines joining the united organs. In Fig. 151b the petals are united and one stamen is attached to each petal (connation and adnation, respectively, 234). If the petals are united below (at their bases) but free above this is indicated as in Fig. 151d where the sepals are united for the whole of their length and the stamens are united

into a tube. The aestivation (148) of the flower is included in the diagram (for petals, quincuncial in 151a, valvate in 151b, contorted in 151d) and the alignment of members of one whorl to another is shown. If the imaginary radius that passes through the centre of a petal also passes through the gap or join between two sepals, the petals are said to alternate with the sepals (151d). If the members of adjacent whorls are on the same radius they are said to be opposite. The stamens in Fig. 151b are opposite to the petals.

These are confusing terms as they have completely contrary meaning when applied to leaf arrangement, phyllotaxis (218). The flowers of many species have fewer members in some whorls than might be expected from theoretical considerations. Missing organs can be included in the diagram by means of a dot or asterisk (151e). The degree of symmetry in a flower is accentuated by its floral diagram. In a floral diagram the different members of a whorl are typically shown all the same size and thus no attempt is made to indicate the considerable difference in petal size and shape often typical of zygomorphic flowers. This aspect of floral morphology is contained in accurate half-flower diagrams, the flower being depicted as if cut through in exactly the median plane and viewed from the cut side (151c). The floral diagram layout can be used to describe shoot complexes (8) and the 'flower' of the Euphorbiaceae (144, 151f). Also the half-flower and floral diagram can be supplemented by adding a floral formula.

This is a code indicating the number of flower parts, and whorls, the attachment of parts, and the nature of the gynoecium. The floral formula for *Lamium album* (151c, e) is

$$\cdot | \cdot \ K(5) \ \widehat{C(5)} \ A4 \ \underline{G(2)}$$

where $\cdot | \cdot$ indicates a zygomorphic flower ($\oplus$ = actinomorphic and $\circledcirc$ = spiral not whorled parts), K = calyx, C = corolla, A = androecium, and G = gynoecium. Numbers refer to the number of members in a whorl, e.g. 5 calyx, and brackets that they are united together. A very large number is shown by ∞, and 0 if absent. If members of two separate whorls are joined, this is indicated by bridging lines, or square brackets:

$$\cdot | \cdot \ K(5) \ [C(5) \ A4] \ \underline{G(2)}$$

A bar above or below the gynoecium number indicates an inferior or superior ovary respectively, this feature not being shown in a floral diagram.

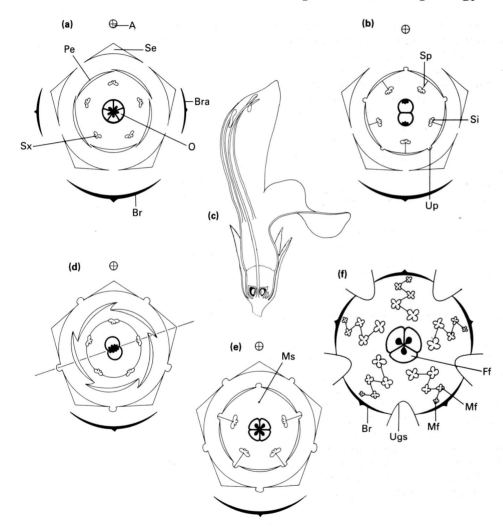

Fig. 151. Floral diagrams. a) symbols; b) adnation, connation; c) *Lamium album*, half flower; d) petals alternate with sepals; e) floral diagram, *Lamium album*; f) floral diagram of terminal cyathium (144) of *Euphorbia* sp. A: axis. Br: bract. Bra: bracteole (62). Ff: female flower. Mf: male flower (stamen only). Ms: missing stamen. O: ovary. Pe: petal. Se: sepal. Si: introrse stamen. Sp: sepal adnate to stamen. Sx: extrorse stamen. Ugs: united glandular stipules of bract. Up: petal connate to petal.

Pollination takes place when pollen grains liberated from an anther (146) are transferred to the stigmatic surface of the same flower, or of a different flower on the same plant, or to a flower on a different plant. Every flower shows specific morphological features which function in this transfer, and a great deal of detail has been published on this topic (e.g. Darwin 1884; Knuth 1906; Proctor and Yeo 1973; Faegri and PiJl 1979). Pollination mechanisms, especially those involving insects, often incorporate the rapid movements of parts of the flower which are difficult to observe. In self-pollinated flowers the slow repositioning of parts may take place and again require careful study. Classifications of pollination mechanisms are based on varying criteria such as: (a) the agency of pollination, e.g. wind, water, invertebrate (bee, butterfly, ant, beetle, fly, wasp, and mollusc), vertebrate (bat, bird, mammal, reptile) and also self-fertilization. (b) Mode of attraction, e.g. colour, shape, smell, movement, nature of pollen, nectar, or lack of attraction. A third system (c) relies more closely on the morphological features of the flower rather than the agencies or attractants (Faegri and Pijl 1979), e.g.

Fig. 152a. *Aristolochia tricaudata*; Fig. 152b. *Aristolochia trilobata*
The perianth segments are united into a tube in which insects are trapped until pollination has taken place.

(1) Flower opening when shedding pollen
   (a) flower inconspicuous (**213b**)
   (b) Flower conspicuous
      i. dish- or bowl-shaped (**153a**)
      ii. bell- or funnel-shaped (**153b**)
      iii. head- or brush-shaped (**153g**)
      iv. gullet-shaped (**153d**)
      v. flag-shaped (**153h, i**)
      vi. tube-shaped (**153f, j**)

(2) Flower opened by visitor when shedding pollen (**153c, e**)

(3) Flower forming a trap for visitors (**152a, b**)

(4) Flowers that are permanently closed and inevitably self-pollinated

(5) Flowers that have a self-pollination mechanism in addition to other mechanisms.

This system applies equally to a whole inflorescence operating as a single floral unit (pseudanthium), such as the capitulum (**141j, 144**) of a member of the Compositae. Any number of similar systems could be devised or refined.

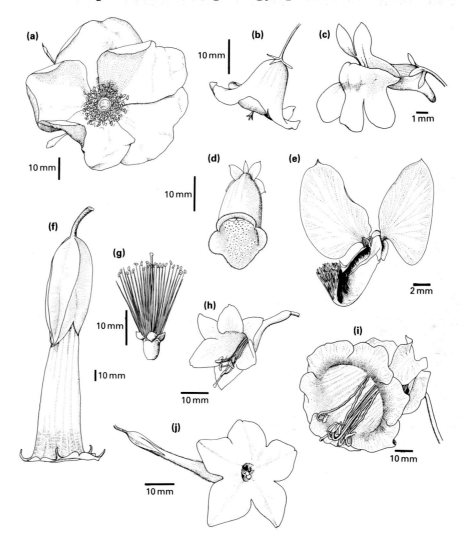

Fig. 153. a) *Rosa rugosa*, dish; b) *Campanula persicifolia*, bell; c) *Cymbalaria muralis*, opening mechanism; d) *Digitalis purpurea*, gullet; e) *Polygala virgata*, opening mechanism; f) *Datura sanguinea* (cf. **235d**), tube; g) *Callistemon* sp., brush; h) *Hosta sieboldiana*, flag; i) *Cobaea scandens*, flag; j) *Nicotiana tabacum*, tube.

Fig. 154. *Sterculia platyfoliacia*
The five carpels of each syncarpous flower (**147h**) separate at an early stage of development and each subsequently dehisces in the manner of a follicle (**157q**).

Botanically speaking, the term 'fruit' is applied to a structure bearing or containing seeds (i.e. fertilized ovules 146) regardless of its edible qualities. The fruit may develop from the ovary (146) of a single flower and then if it represents the single carpel of the flower or the single ovary of a syncarpous flower (**147h**) it is called a simple fruit (**157a–h, o–v**). If the fruit is derived from one apocarpous flower, in which the carpels are not united (**147g**), it is referred to as an aggregate fruit or etaerio (**157i–k**). A fruit formed from a group of flowers is termed a multiple fruit (**157l–n**). Structures other than the gynoecium, such as the receptacle (146) or perianth members, are sometimes incorporated into the formation of a fruit and such composite structures are termed false fruits, or pseudocarps, or accessory fruits (**157g, h**). The basis of classification of fruit types is partly morphological and partly related to the mechanism of seed dispersal (160) and is not particularly logical. A representative summary (Pijl 1969) of the most common terms used in floras is given here (156); fruits are traditionally identified as being either 'dry' or 'fleshy' and as being either 'dehiscent' (breaking open to release seeds) or 'indehiscent' (remaining closed). Fleshy fruits often contain a hard woody layer, representing part of the fruit wall or pericarp. The pericarp is derived from the ovary wall and is composed of three layers, an outer epicarp, and an inner endocarp, with a mesocarp between. Thus the outer surface of a coconut (*Cocos nucifera*) or peach (*Prunus perisica*) is the epicarp, the coconut fibre or peach flesh is the mesocarp, and the hard shell or stone the endocarp. Both are technically drupes (**157f**) and the thin brown layer immediately within the hard endocarp is the seed coat (testa) surrounding the embryo (plus or minus endosperm 163). In a berry (**157e**) the testa is woody, all layers of the pericarp being fibrous or fleshy. Thus the outer pigmented skin (or 'zest') of a *Citrus* fruit is the epicarp, the white layer beneath is the mesocarp, and the endocarp consists of a mass of succulent hairs, the edible portion. Each seed has a hard testa. This distinctive type of berry, having a hairy endocarp, is termed an hesperidium. A typical berry (**157e**) is represented by a grape (*Vitis vinifera*).

(Continued on page 156.)

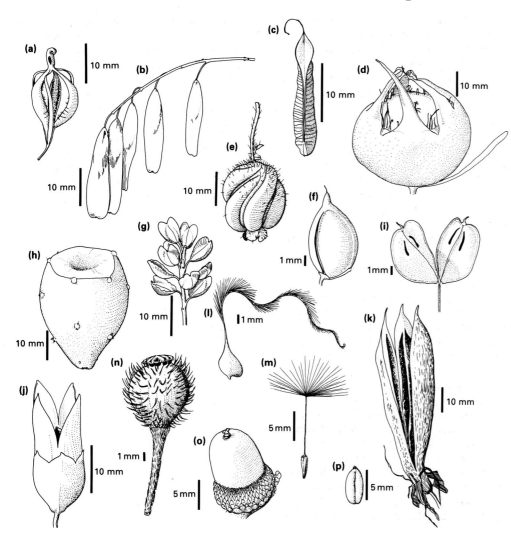

Fig. 155. a) *Epidendrum* sp., septifragal capsule; b) *Fraxinus excelsior,* samara; c) *Aquilegia vulgaris,* follicle; d) *Mespilus germanica,* pome; e) *Blumenbachia insignis,* septicidal capsule; f) *Carmichaelia australis,* legume; g) *Phlox* sp., silicle; h) *Opuntia* sp., berry; i) *Heracleum sphondylium,* schizocarp; j) *Vestia lycoides,* capsule; k) *Phormium tenax,* loculicidal capsule; l) *Clematis montana,* achene; m) *Taraxacum officinale,* achene, inferior ovary; n) *Papaver hybridum,* poricidal capsule;, o) *Quercus petraea,* nut; p) *Triticum aestivum,* caryopsis.

Fig. 156. *Entada* sp.
A massive woody legume.
The Alvis 12/50 headlight centres are 600 mm apart.

A. Achene. Indehiscent. Dry. Usually small with one carpel.

B. Caryopsis. Achene in which testa and pericarp are fused (as in Gramineae 186).

C. Nut. Indehiscent. Dry. Woody pericarp. Usually large with more than one carpel.

D. Samara. Winged achene.

E. Berry. Indehiscent. Fleshy pericarp, woody testa (grape *Vitis vinifera*).

F. Drupe. Indehiscent. Fleshy epicarp and mesocarp. Endocarp woody. Testa not woody (plum *Prunus domestica*).

G. Pome. Indehiscent. Fleshy. Pseudocarp. Receptacle fleshy. Testa woody (apple *Malus pumila*).

H. Cupule. Pseudocarp. Bracts incorporated into the fruit construction.

I. Etaerio (aggregate) of achenes.

J. Etaerio (aggregate) of berries.

K. Etaerio (aggregate) of drupes.

L. Strobilus. Dry multiple fruit of achenes incorporating bracts (hop *Humulus lupulus*).

M. Sorosis. Fleshy multiple fruit (mulberry *Morus* spp., pineapple *Ananas comosus* **223**).

N. Syconium. Fleshy multiple fruit with achenes attached to infolded receptacle (**241**).

O. Schizocarp. A fruit that breaks apart without releasing seeds. Each part termed a mericarp or coccus contains one seed and is indehiscent.

P. Lomentum. A schizocarp derived from an atypically indehiscent legume (R).

Q. Follicle. Dehiscent. Single carpel splitting down one side.

R. Legume. Dehiscent. Single carpel splitting down two sides.

S–Y. Capsule. Dehiscent. More than one carpel.

S. Silique. Dehiscent. Two carpels splitting away from central column—the replum.

T. Silicle. Dehiscent. A short silique.

U. Pyxidium (Pyxis). Dehiscent capsule with lid.

V. Poricidal capsule. Dehiscent forming pores.

W. Loculicidal capsule. Dehiscent. Capsule in which each carpel splits open.

X. Septifragal. Dehiscent. Capsule in which the seeds remain attached to a central column. (Can be loculicidal as shown or septicidal Y).

Y. Septicidal. Dehiscent. Capsule in which the carpels separate.

A more extensive list is to be found in Radford *et al.* (1974).

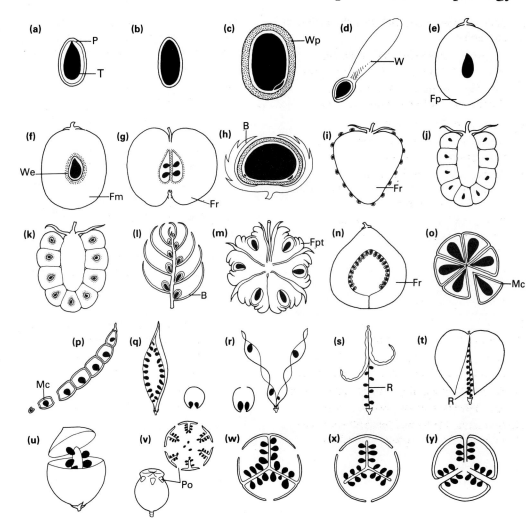

Fig. 157. Fruit types (156) (seeds depicted black). B: bract. Fm: fleshy mesocarp. Fp: fleshy pericarp. Fpt: fleshy perianth. Fr: fleshy receptacle. Mc: mericarp. P: pericarp. Po: pore. R: replum. T: testa. W: wing. We: woody endocarp. Wp: woody pericarp.

Fig. 158. *Paullinia thalictrifolia*
A black shiny seed hangs on a white aril out of an opened, red, berry.

Each ovule inside an ovary (146) develops into a seed when fertilized. The stalk of the ovule and of the subsequent seed is termed the funicle and may play a part in the dispersal of the seed (160) from the mature ovary (the fruit 154). If the seed is detached from the funicle it will leave a scar, the hilum (**159e**). The distal end of an ovule is enveloped by one or two layers of tissue, the integuments, which usually do not meet completely at the top leaving a hole, the micropyle (**159l**), through which the pollen tube may find entry to the ovule at fertilization. As the ovule enlarges into a seed, one or both integuments develop into the seed coat or testa. The micropyle may remain visible on the seed (**159e**). Some ovules are bent over on the funicle (anatropous **159m** as opposed to orthotropous **159l**) and the micropyle in the seed is therefore next to the hilum, the funicle appearing as a ridge fused down the side of the seed and then known as the raphe. The seed coat can be much elaborated and very hard (sclerotesta). If it develops with a soft layer it is termed a sarcotesta. In many seeds dispersed by animals and birds (Pijl 1969) there is a hard testa with a conspicuous swollen fleshy addition (an 'arilloid'). If this outgrowth is on the raphe it is specifically termed a strophiole (**159a**), if next to the micropyle, a caruncle (**159b**) (especially if hard). More elaborate structures at the micropylar end of the seed are termed arillodes (**159c**), which if detached can leave an additional scar referred to as a false hilum. A fleshy outgrowth of the funicle enveloping most of the seed is termed an aril (**158, 159d, f**). The term 'seed' is often applied inaccurately to whole fruits or parts of fruits, particularly when the true seed is fused inside a dry indehiscent pericarp wall. The 'seeds' of grasses (Gramineae), umbellifers (Umbelliferae), and beets (Chenopodiaceae) are morphologically fruits (caryopsis, mericarp, and achene, or nut, or multiple fruit, respectively 156).

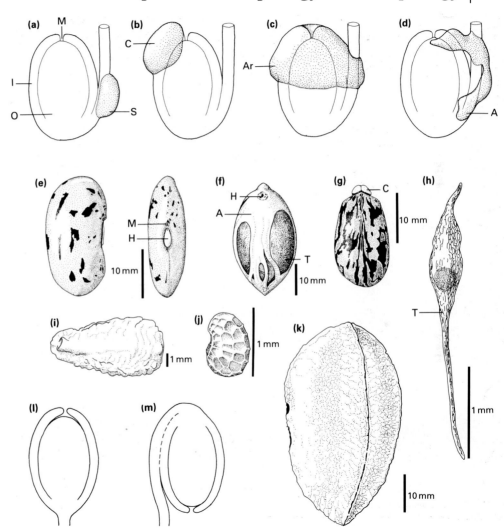

Fig. 159. a)–d) arilloids, a) strophiole, b) caruncle, c) arillode, d) aril. e)–k) single seeds, e) *Phaseolus vulgaris*, f) *Myristica fragans*, g) *Ricinus zanzibarensis*, h) *Epidendrum ibaguense*, i) Proboscidea louisianica, j) *Papaver hybridum*, k) *Bertholletia excelsa*, l) orthotropous ovule, m) anatropous ovule. A: aril. Ar: arillode. C: caruncle. F: funicle. H: hilum. I: integument. M: micropyle. O: ovule. S: strophiole. T: testa.

Fig. 160. *Geranium* sp.
Two fruits. Each has exploded. Five carpels are sprung out
on strips of curling style (called awns) and the seeds have
been catapulted out.

Plants are dispersed by the release of detachable
portions that can become established away from
the parent. Such structures—fruits, seeds, bulbils
(172) are known as diaspores, a unit of dispersal.
Fruits as a whole, or individual seeds, are
dispersed by a range of agents of which the most
frequently cited are wind, water, and animal
(bird, insect, etc.) or by ballistic mechanisms.
Both fruits and seeds exhibit morphological
constructions that can be shown to operate
during dispersal; long hairs on a wind-borne
achene for example (3, 155l, m). An elaborate
but accurate terminology exists to classify the
different modes of dispersal (Pijl 1969) (e.g.
myrmecochory, by ants; epizoochory, diaspore
detached from the plant and attached by some
mechanism to an animal). Wind dispersed
fruits/seeds usually have some structure
increasing surface area. This may take the form
of a wing (155b), or parachute (155m, 161f), or
hairs (155l). Hooked spines (usually emergences
76) on a fruit usually indicate dispersal on fur or
feathers (161b, e); release of the fruit may
involve a passive 'shaking' ballistic mechanism
(161d). Passive ballistics also occur in plants
with poricidal capsules (155n). Active ballistic
mechanisms involve sudden rupture of dehiscent
fruit on drying out (160) or mechanisms based
on increase in turgidity (161c). Fruits and seeds
eaten and therefore dispersed by animals and
birds are typically fleshy in part (161g). The
edible tissue may form part or all of the fruit wall
(pericarp), the endocarp of which may intrude
into the ovary cavity between the seeds (the
pulpa). If the fruit itself is not fleshy and

attractive, then it may dehisce and expose contrasting colours of fruit, seed, and seed appendages, e.g. aril (**158**). The seed stalk (funicle) may be coloured or fleshy or elongate, dangling the seeds out of the fruit. The seed coat itself is usually hard except in the case of sarcotesta (**158**). Fruits or seeds dispersed (or rather collected) by ants bear an oil secretion structure, the elaiosome, which may represent a modified caruncle or strophiole on a seed (**159**), or various structures on fruits or multiple fruits. Fruits and seeds may not be dispersed at all, but simply drop around the plant (**155o**) or even be placed in or on the soil by bending pedicels (**267a, c**). If a seed germinates in the fruit before the fruit is detached from the plant, the plant is said to be viviparous (**166**).

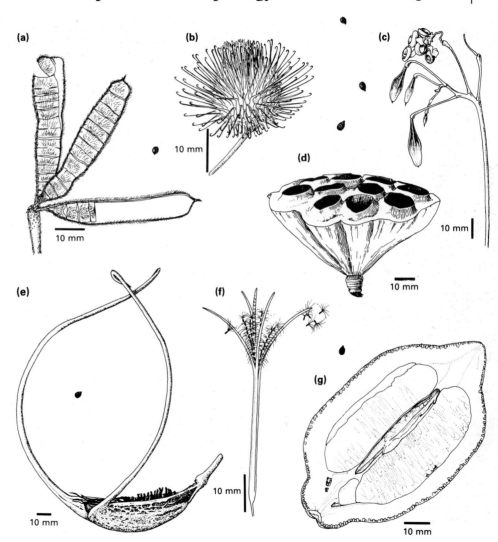

Fig. 161. a) *Mimosa berlondiera*, passive ballistics; b) *Arctium minus*, animal, hooked bracts; c) *Impatiens glandulifera*, active ballistics; d) *Nelumbo nucifera*, passive ballistics plus water; e) *Proboscidea louisianica*, animal, hooked fruit; f) *Epilobium montanum*, wind, seeds plumed; g) *Citrus limon*, animal, active, fleshy endocarp.

Fig. 162. *Ocimum basilicum*
Seedlings from above. Epigeal germination (cf. Fig. **165a**)
with pairs of relatively large cotyledons flanking pairs of
developing foliage leaves.

A seed (158) usually contains just one embryo in which the first stages of differentiation of tissues and organs have located the potential shoot system and root system. The seed also contains a limited quantity of stored food which will allow the embyro to grow out of the seed coat and develop into an independent photosynthesizing, i.e. food producing, structure. This is the process of germination (164), and the young plant is termed a seedling up to an indeterminate arbitrary age (establishment 168, 314). The morphological details of the embryo within the seed and of the seedling as it emerges, vary depending upon the type of germination and the nature of the plant, dicotyledon or monocotyledon. The embryo and therefore the seedling of a dicotyledon possesses two leaves (cotyledons) attached at the cotyledonary node of an axis that has a primary root (or radicle) at one end and a shoot apical meristem (16) or plumule at the other (**163**). The junction of the root end and the shoot end (called the transition zone) can be more or less abrupt and not necessarily easily identifiable without anatomical investigation of the vascular system, although a prominent 'root collar' or 'peg' (**163c**) may be present at this point. The portion of axis between the cotyledonary node and the transition zone is called the hypocotyl (166), that immediately above the cotyledons, the epicotyl. The cotyledons themselves may be variously shaped, lobed or elongated, and fold together within the seed in numerous ways similar to the folding of the leaves in a bud (38). Usually they are of similar size and shape but in some dicotyledons one is very much larger than the other (**163f, 209**). They are usually opposite each other at the node; the location of subsequent leaves on the stem progressively conforms to the phyllotaxis of the mature shoot (**218**). Axillary buds, sometimes more than one (**236**), occur in the axil of each cotyledon, which often has a very short petiole. Cotyledons play a crucial role during the process of germination (**164**) and may contain stored food, or become photosynthetic, or both. The single cotyledon of the monocotyledon does not contain stored food, this being present in the form of endosperm adjacent to the embryo inside the seed. The cotyledon absorbs the endosperm when the seed germinates. It may also perform a photosynthetic function (**164**).

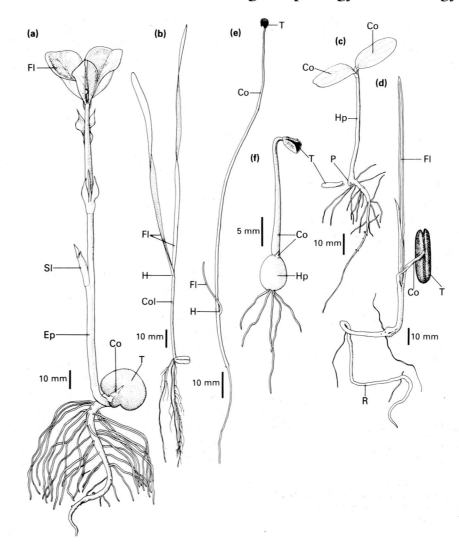

Fig. 163. a) *Vicia faba*, hypogeal dicotyledon; b) *Triticum aestivum*, hypogeal monocotyledon; c) *Cucumis sativus*, epigeal dicotyledon; d) *Phoenix dactylifera*, hypogeal monocotyledon; e) *Allium cepa*, epigeal monocotyledon; f) *Cyclamen persicum*, epigeal dicotyledon (anisocotyly 32). Co: cotyledon. Col: coleoptile (164). Ep: epicotyl. Fl: foliage leaf. H: hole. Hp: hypocotyl. P: peg. Sl: scale leaf. T: testa.

During germination, morphological development transforms a seedling dependent upon food stored in the seed, into a seedling able to photosynthesize its own food. At first water is taken into the seed by imbibition, but the first act of establishment (168) for the seedling is the production of a root or roots that will absorb additional water and anchor the plant. The food stored in the seed is either present in the cotyledons (dicotyledons only), and/or as endosperm, a product of fertilization in addition to the embryo and situated within the testa alongside the embryo. In some species part of the ovule tissue (154) also acts as a food source, the perisperm. A number of different sequences of development at germination can be recognized (165a–g) depending upon the food source, the role played by the cotyledon(s) and the manner in which the seedling axis elongates producing a photosynthetic array of leaves. Two principle modes of germination are given the names epigeal and hypogeal with reference to the location of the cotyledon(s) during this process; 'geal' indicates soil surface ('earth'), 'epi' above, and 'hypo' below. Thus an epigeal seedling (165a, c, e) develops such that its cotyledons (163c) or cotyledon (163e) is above ground. During hypogeal germination, axis elongation is such that the cotyledons (165b) or cotyledon (165f, g) remain below ground or at least at ground level. For the cotyledons to be carried above ground, the portion of axis beneath the cotyledons (the hypocotyl 166) must elongate. For the cotyledons to remain below ground, the portion of axis above the cotyledons (epicotyl)

must elongate, i.e. hypogeal germination, epicotyl elongates; epigeal germination, hypocotyl elongates. The permutations of functions performed by the cotyledon(s) in germination is illustrated in Fig. 165. An unusual form of germination in which elongation of the cotyledonary petioles plays a role is found in *Vitellaria paradoxa* (41g).

In monocotyledons the cotyledon may elongate at germination, its distal tip remaining within the seed coat with the endosperm and its proximal end pushing the rest of embryo out of the seed coat (163d, 165f, g). The cotyledon is attached to the stem axis around most or all of its circumference as is typical of monocotyledon leaves (14), and can thus form a tube or solid structure with hollow base with the shoot apex at first hidden inside. The second leaf then emerges through a hole or slit in the side of the cotyledon (163e). In other monocotyledons the cotyledon does not elongate (grasses) but remains within the seed absorbing endosperm. However, different interpretations are applied to the exact extent of the cotyledon in these plants; the second structure produced, which forms a photosynthetic sheath, the coleoptile, is regarded either as the second leaf or as part of the cotyledon itself (180). The primary root (radicle) of a dicotyledon seedling will bear lateral roots as it enlarges in circumference. A number of root primordia (94) are usually present in a monocotyledon embryo in addition to the primary root. These are adventitious roots (98) associated with leaf nodes. Roots present as primordia in the embryo before germination are

referred to as seminal roots. A distinction is made, at least in grasses, between the primary seminal root (radicle) and the lateral seminal roots (180).

Fig. 164. *Cucurbita pepo*
Seedling with epigeal germination. The photosynthetic cotyledons have expanded considerably in size and are now much larger than the seed coat (testa) (still visible) that contained them.

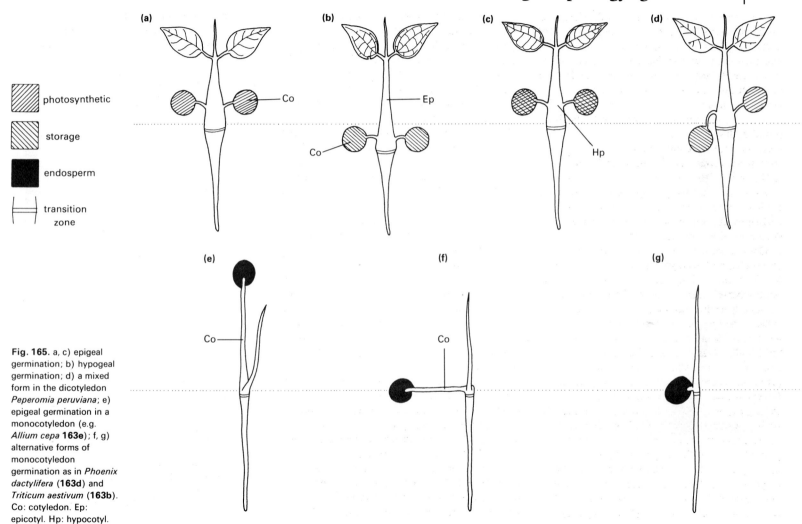

Fig. 165. a, c) epigeal germination; b) hypogeal germination; d) a mixed form in the dicotyledon *Peperomia peruviana*; e) epigeal germination in a monocotyledon (e.g. *Allium cepa* **163e**); f, g) alternative forms of monocotyledon germination as in *Phoenix dactylifera* (**163d**) and *Triticum aestivum* (**163b**). Co: cotyledon. Ep: epicotyl. Hp: hypocotyl.

photosynthetic

storage

endosperm

transition zone

Fig. 166. *Rhizophora mangle*
The embryo germinates whilst the fruit and seed are still attached to the plant. Germination results in the elongation of the hypocotyl which constitutes most of the seedling at this stage. The epicotyl is still inside the seed, the radicle (seedling root) is at the lower pointed end of the seedling.

The hypocotyl is the length of stem linking the node at which cotyledons (in dicotyledons) are attached to the proximal end of the primary root. It is an elongated structure in plants with epigeal germination (**163c**). The junction of hypocotyl and root, the transition zone, is often ill-defined externally and is anatomically distinctive internally. The vascular anatomy of the hypocotyl is dominated by the veins serving the cotyledons. The hypocotyl shows some root-like features, and may bear 'root' hairs and often adventitious roots (98). Having no leaves by definition, it can therefore superficially resemble the primary root with which it is directly connected. It also frequently bears buds, these being termed adventitious (**167e**) as they are not in leaf axils. Thus an extensive shoot system can arise below the cotyledons, and the plumule (162) may fail to develop. The hypocotyl can be contractile in exactly the same manner as a contractile root (106). It can also form a storage tuber and become considerably swollen. In different species the stem internodes above the cotyledons and the root below the transition zone may or may not be swollen also. Without a developmental study of the anatomy of these tubers it is difficult to judge how much represents stem or hypocotyl or root (**167a–d**). The stem portion should bear leaves or leaf scars and the root portion lateral roots, possibly in orderly vertical rows. Elongation of the embryo hypocotyl before seed and fruit are shed gives rise to vivipary, in some mangroves for example (**166**), the whole seedling eventually dropping from the tree. True vivipary, the germination of

seed before dispersal, is relatively rare, in contrast to false vivipary (proliferation 176) which represents a development of rooting vegetative buds instead of flowers. The hypocotyl of very young seedlings of some species of parasitic plants plays a part in the attachment of the parasite to the host (108).

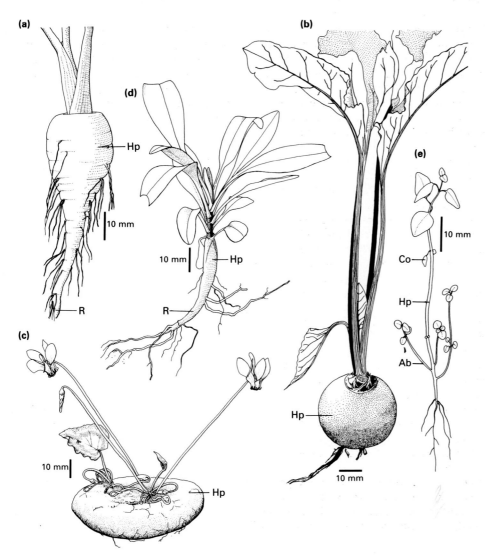

Fig. 167. a–d) root tubers (110) incorporating an upper (proximal) portion of swollen hypocotyl. a) *Pastinaca sativa*, b) *Beta vulgaris*, c) *Cyclamen hederifolium*, d) *Centranthus ruber*, e) *Antirrhinum majus*, Ab: shoot formed from adventitious bud (232). Co: cotyledon. Hp: hypocotyl. R: radicle.

The process of germination (164) establishes a young seedling such that it is anchored in the ground, can take up water, and can photosynthesize. The process of establishment continues however, and is marked by a sequence of morphological events elaborating the root system and extending the shoot system. The morphological status of the plant (314) may bear no relation to its actual age; very small tree 'seedlings' in a forest may grow millimetres a year and be of considerable age before conditions allow substantial increase in size. Establishment is a particularly notable aspect of monocotyledonous seedling development. Stems and roots of these plants mostly lack the ability to grow in girth, all roots are relatively thin and borne on the stem (adventitious roots 98), and an increase in stem surface at ground level thus precedes extra root production (Holttum 1954). A number of modes of establishment growth in monocotyledons are noted in Tomlinson and Esler (1973) as follows:

A. (**169c**) Each successively produced internode of a palm seedling axis is slightly wider than the previous one. The internodes themselves are very short and the result is that the seedling develops in the form of an inverted cone which is kept buried in the soil by contractile roots (106). Once the cone is established, a trunk can develop by the production of longer internodes, and the large cone surface has room for many adventitious roots.

B. (**169i**) Many monocotyledons are rhizomatous (130) and with very few exceptions the rhizome develops sympodially (250). This is also true of the seedling establishment—buds at the base of the plumule grow out as small stout sympodial units, the distal ends of which grow erect; these in turn bear buds giving rise to slightly larger units, and so on. The seedling becomes established in terms of size, spread, and stem surface next to the ground allowing adventitious root development.

C. (**169h**) A variety of the sympodial sequence of B involving change of growth direction with respect to gravity leads to even greater rooting surface and firm planting of the established seedling.

D. (**169f**) This change of orientation may be confined to the seedling plumule alone, with no lateral branches but with a pronounced increase in stem width from internode to internode as in A.

E. In some *Cordyline* species (Agavaceae) the plumule grows in the manner of A above but not necessarily with such compact internodes, and can increase in girth due to secondary thickening (16). In addition, one bud near the base of the seedling develops into a rhizome that grows vertically downwards, anchoring the plant on production of adventitious roots (**169k**). A sequence of such orientation changes is exhibited by *Costus spectabilis* (**169d**).

F. (**169g**) Seedling growth may be rapid and orthotropic (246), stability being maintained by a climbing or scrambling habit accompanied by the production of prop roots (102).

These forms of establishment are not necessarily confined to the seedlings; dormant buds developing much later (reiteration 298) may produce shoot sequences similar to that of the original seedling axis. The seedling axis of a dicotyledonous plant is able to increase in size indefinitely owing to cambial activity (16, **169e**) in most instances and the process of establishment does not usually take the same form as that of monocotyledons, the root system being largely derived from branching of the radicle rather than by adventitious root production. In dicotyledonous plants with rhizomes or stolons (**169a**), the role of the radicle is not so pronounced, lateral spread of these stems giving added shoot to ground contact and potential for adventitious root production as in monocotyledons. Radial production of branches from a seedling establishes a pattern that is not necessarily maintained by later branching sequences (306, 312). Nevertheless, seedling dicotyledons can possess contractile roots and contractile hypocotyls, and elaborate establishment mechanisms exist such as that of *Oxalis hirta* (**169j**). Precocious development of buds in the axil of a cotyledon, or buds on the hypocotyl (**167e**), also act as establishment mechanisms and bending of axes can play a part as in *Salix repens* (**267b**). An extreme form of precocious development is found in plants in which the seedling germinates while still contained in the fruit, and whilst the fruit is still attached to the parent plant. Such precocious development is termed vivipary, an example of which occurs in the mangrove *Rhizophora* (166). An account of the range of establishment types of forest trees in relation to their mode of germination is given by Miquel (1987), and of plants with underground storage organs (170) by Pate and Dixon (1982).

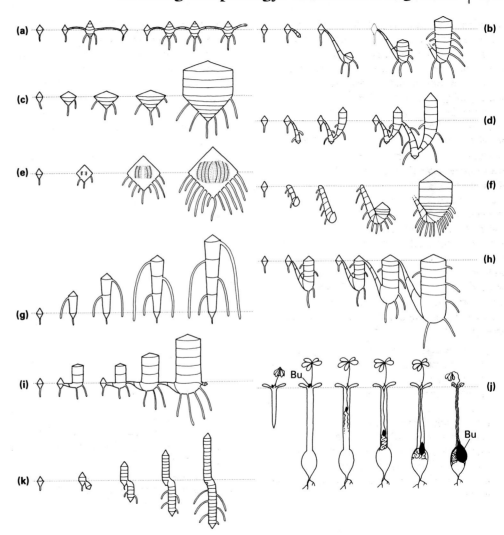

Fig. 169. Examples of establishment growth. a) production of stolons from parent plant; b) production of successively larger, but short-lived sympodial units (compare h); c) increase in width of successive internodes; d) increase in size of successive sympodial units with alternating growth direction; e) increase in width due to cambial activity; f) as 'c' with initial downward growth; g) initial vertical growth supported by prop roots; h) increase in size and depth of successive long-lived sympodial units; i) increase in size of successive sympodial units; j) *Oxalis hirta*, contraction of radicle with elongation of foliage leaf petiole resulting in descent of its subtended bud (Bu) (Davey 1946); k) production of single downward growing side shoot.

Vegetative multiplication (also referred to as vegetative reproduction if contrasted with sexual reproduction) is a process involving the death of tissues located such that part of an existing plant becomes detached and independently rooted. Precise abscission zones may form as in the detachment of bulbils (172) or a plant may fragment due to decay between living components as in the case of root buds (178). Each of the morphological structures known as rhizome (130), stolon (132), runner (134), corm (136), bulb (84), and tuber (stem 138, root 110) undergo vegetative multiplication by death and decay of old tissue. Stolons and runners consist of relatively long and thin sections of stem having long internodes, alternating with sections with very short internodes, and producing adventitious roots (98). Death of the stolon or runner separates these rooted and now independent daughter plants (each of which is termed a ramet, collectively a genet or clone) (171a, b). A rhizome is typically a stouter stem than a stolon and usually fragments only if it is branched, the old proximal portion decaying and separating the ramet into two new ramets each time the rotting reaches a branch junction (171c, d). Definitions of 'rhizome' usually emphasize horizontal growth below ground level. A number of epiphytic plants have stems with a rhizome morphology growing more or less vertically on tree trunks (294a). Some species of woody monocotyledon (e.g. *Cordyline*) produce 'aerial' rhizomes developing in a downward direction similar to the seedling establishment rhizome of the same plant (169k), and able to

root as independent plants in some circumstances. A corm consists of a squat swollen stem orientated vertically in the soil and bearing daughter corms, sometimes called cormels, at its distal or proximal end. The daughter corms represent buds developing in the axils of leaves on the parent corm. Eventual death of the parent corm separates the daughters (171f, h). The same procedure of vegetative multiplication occurs in bulbs (171m). Again the stem axis is vertical but food is stored in leaf bases or scale leaves rather than in the swollen stem as in the case of a corm. Buds in the leaf axils may develop into daughter bulbs which will be located in a radial manner around the parent, the latter eventually rotting away. A tuber may be formed from a swollen stem bearing buds in leaf axils or a swollen root bearing adventitious buds not associated with leaves. In each case the tuber often has a narrow, possibly elongated, connection to the parent plant and breakage or rotting of this connection results in vegetative multiplication (171k). Long thin underground stems connecting stem tubers to the parent plant are sometimes referred to as 'stolons'. Root tubers are connected to the parent by thin roots. Some stem tubers are sympodially branched to a limited extent and disintegrate into a corresponding number of parts. They may be regarded as rhizomes or tubers depending upon the definition employed. Intermediate types will always be found in any attempt to categorize sharply morphological structures (e.g. 296). Aerial shoots of some herbs and shrubs bend under their own weight, touching the soil.

Adventitious root production (98) then results in natural layering, the rooted portion of the stem becoming an independent plant if connections with the parent decay. An extensive account of underground storage organs is given in Pate and Dixon (1982).

Fig. 170. *Cylindropuntia leptocaulis*
The large round fruits bear detachable propagules with tenacious barbed spines.

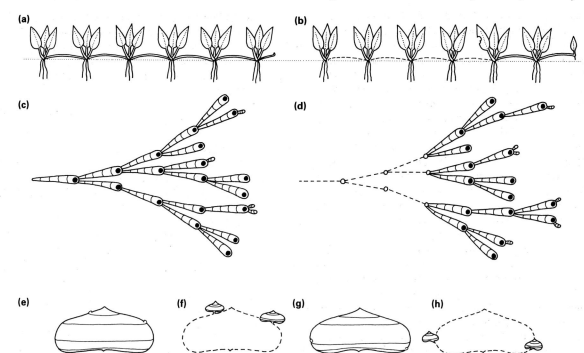

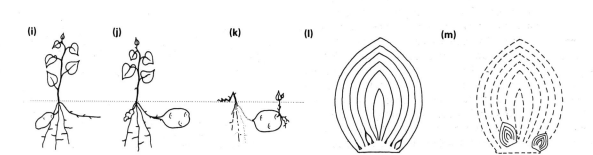

Fig. 171. Examples of natural cloning. a, b) separation of individual ramets in stoloniferous plant by death of intervening connections; c, d) disintegration of rhizomatous plant, plan view, by death of proximal components; e–h) production of daughter corms on parent corm which subsequently rots; i–k) persistence of stem tuber after death of remaining plant; l, m) production of daughter bulbs in parent bulb which subsequently rots.

Fig. 172. *Lilium* cv. minos
A detachable bulbil with adventitious roots has formed in place of a flower.

A bulbil is merely a small bulb (84), that is a short stem axis bearing fleshy scale leaves or leaf bases and readily producing adventitious roots. However, the term is also sometimes inaccurately applied to any small organs of vegetative multiplication such as axillary stem tubers found on the aerial stems of some climbers (**139a**). Also there are alternative terms, bulblet, bulbet, bulbel, which are variously given precise definitions or used indiscriminately as synonyms. Small bulbs are mostly found in one of two locations, either on an aerial stem, representing axillary buds and especially replacing flowers in an inflorescence (176), or developing in the axils of the leaves of a fully sized bulb. The former are the type most consistently termed bulbils. Small bulbs developing within an existing bulb are of two types: one or more larger bulbs which will replace the parent bulb (renewal bulb in the terminology Mann 1960) and a number of smaller bulbs in scale leaf axils which will be liberated on the death of the parent bulb (increase bulbs). Often more than one increase bulb is produced in the axil of each leaf (**84, 236**) and sometimes they are adnate (234) to the underside (abaxial surface) of the next youngest leaf, or adnate to the adaxial surface of the leaf some distance out from the node. Increase bulbs can be formed on the end of thin stems (dropper 174) and are then dispersed away from the parent bulb. More elaborate mechanisms occur in some plants, e.g. *Oxalis cernua* (**169j**). Very many water plants undergo vegetative multiplication surviving periods of cold, dryness, or nutrient depletion by the production of detachable buds.

These are called turions ('winter buds') regardless of the variety of their form. In the Lemnaceae (212) they simply represent particularly small fronds. In *Utricularia* species the apical meristem produces scale leaves around a compact bud covered with hairs and mucilage. In other plants the turions may more closely resemble other vegetative buds on the plant but are more compact and a darker green. Either lateral buds or the apical bud of a shoot or both may form turions which may be easily detachable owing to the formation of an abscission layer of cells at their base, or may merely persist when the rest of the plant rots. The leaves of a turion contain stored food and adventitious roots are produced when favourable conditions return. This form of 'over-wintering' typifies the category of plant form recorded as hydrophyte by Raunkiaer (1934) (**315g**).

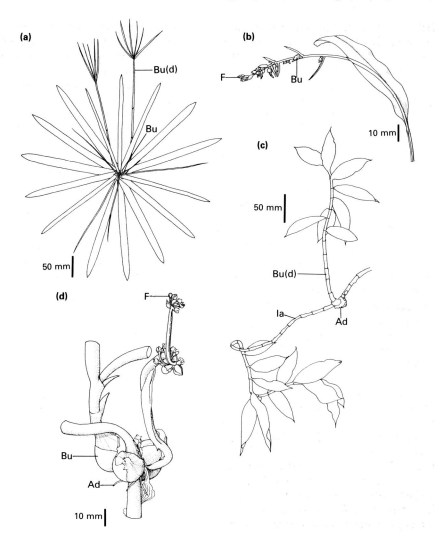

Fig. 173. a) *Cyperus alternifolius*, top of aerial shoot viewed from above; b–d) distal ends of inflorescence axes: b) *Globba propinqua*, c) *Costus spiralis*, d) *Allium cepa* var. *viviparum*. Ad: adventitious root. Bu: bulbil. Bu(d): developing bulbil. F: flower. Ia: inflorescence axis.

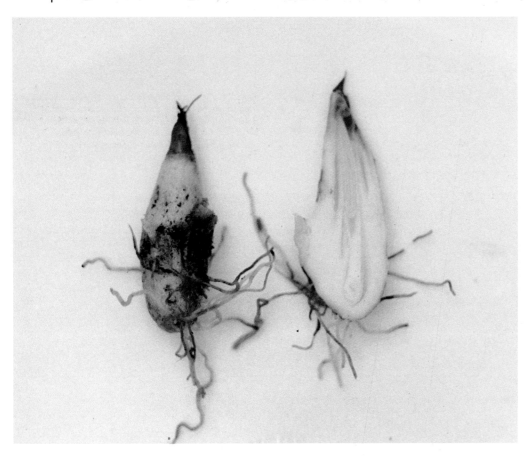

Fig. 174. *Erythronium dens-canis*
The distal end of a dropper. Longitudinal section on right.
Two potential shoots are present. The one in the centre
represents the original bud associated with the dropper. The
lateral bud at the left is in the axil of a scale leaf, now
detached.

Buds produced in the axils of leaves at the base of
a bulb (84, **171m**) or corm (136, **171f, h**)
usually develop into independent plants with
adventitious roots (98) and thus will be located
close to the base of the parent. In a number of
species, however, the bud may be carried away
horizontally or vertically from the parent on the
end of a slender root-like structure (**175h**). These
are referred to as droppers or sinkers. The
sequence of development of a dropper in different
plants varies in detail and can only be deduced
by careful morphological dissection at all stages
of growth. The elongated portion may represent
one very long internode or more accurately
hypopodium (262), which is the portion of axis
between the first leaf (prophyll 66) of the axillary
bud and the parent axis. In some cases an
adventitious root primordium forms adjacent to
the bud and the two grow out as one combined
structure (i.e. they are adnate **175a–e**). The base
of the subtending leaf can also grow out to keep
pace with the developing dropper and form an
outer layer of the whole structure to which it
may be fused (**174**). Enlarged structures formed
in this way are sometimes referred to as stem
tubers (138). Finally the root portion may be
very much larger than the shoot, the swollen
organ then being referred to as a form of root
tuber (110).

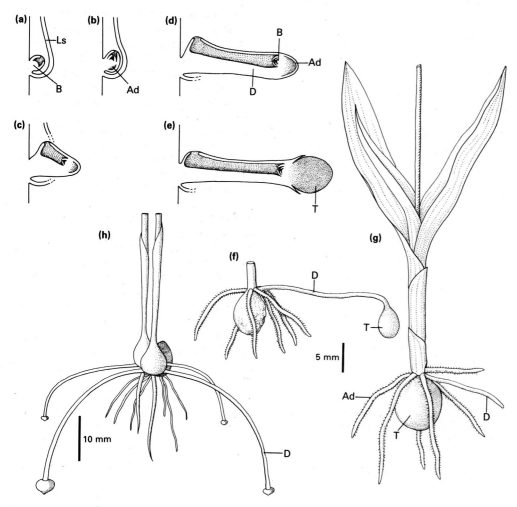

Fig. 175. a–g) Development of a dropper in *Herminium monorchis*; h) *Ixia conica*, bulb with emerging droppers. Ad: adventitious root. B: bud. D: dropper. Ls: Leaf sheath. T: tuber. (a–e, redrawn from Troll 1943; f–h, redrawn from Raunkiaer 1934).

Fig. 176. *Deschampsia alpina*

Small tillers (182) have developed in place of spikelets in the inflorescence. This is false vivipary; true vivipary being the condition in which seeds germinate before being shed from the parent plant (168).

An apical meristem in an inflorescence will develop to give rise to either a flower, or another unit of the branching structure (142). Each such meristem is in the axil of a modified leaf, a bract (62), although these may be absent. Bracts may be present that do not subtend an active axillary meristem. i.e. the bract is 'sterile' (e.g. grasses 186). In some plants, meristems that would normally develop into flowers develop instead into vegetative buds usually associated with adventitious roots (**177a**). These vegetatively produced plants will grow independently if shed or deposited on the ground by the collapse of the inflorescence axis. The leaf bases of the buds may be swollen, the structure resembling a bulb and such a structure is then called a bulbil (172). The production of vegetative buds instead of flowers, or as happens in some grasses, the production of tillers in normally sterile spikelets (**177b, c, d**), is termed prolification or false vivipary (proliferation is also used in this context but cf. 238). True vivipary occurs when a seed germinates without being shed from the parent plant (168). An inflorescence axis will revert to vegetative extension in some plants (**261b**), or axillary shoots may develop that are vegetative rather than reproductive (**199a** and **253a**).

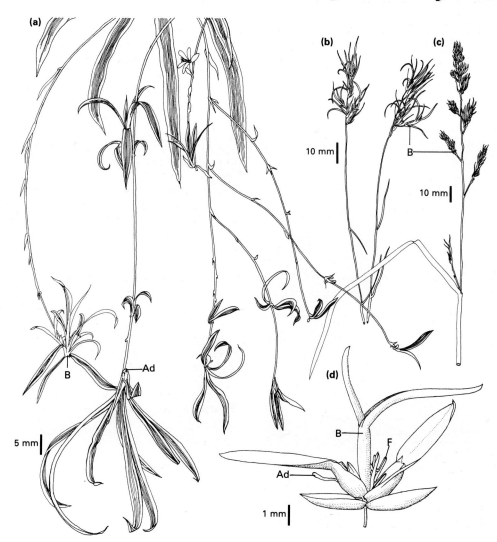

Fig. 177. a) *Chlorophytum comosum*, trailing inflorescence; b) *Festuca ovina* var. *vivipara*, inflorescences with tillers (182) instead of spikelets; c) *Dactylis glomerata*, ditto, cf. **185g**; d) *Poa × jemtlandica*, a single spikelet containing both flowers and vegetative bud. Ad: adventitious root. B: vegetative bud. F: flower.

Buds capable of developing into shoot systems occur on the roots of a number of plant species, both monocotyledons and dicotyledons. In many plants such buds, which are termed root buds, only form if the root is damaged, buds then differentiating in the callus tissue that forms. In a few plants a root bud can form so close to the root apex that without anatomical study of its development it appears as if the root has turned into a shoot. This is particularly so if the root apex itself becomes parenchymatized (244). Root bud primordia arise endogenously (94, **178**), that is within the root tissue as do lateral root primordia, and not exogenously at the surface as is typical of buds arising on stems. The precise location within the root is variable. It may be in exactly the same position as would normally be occupied by a lateral root primordium, typically in the pericycle, and then root buds will appear in a number of rows depending upon the detailed root anatomy (**97k**). Frequently a root bud primordium develops in very close proximity to a lateral root before the latter emerges from the main root. Alternatively the root buds differentiate in the cortex and are not associated in any way with lateral root positions. In some trees, root buds develop within the living part of the root bark and may remain dormant for extended periods. Nevertheless extensive clones of trees develop from such root buds and a stand of trees (e.g. *Populus* spp., *Liquidambar* sp.) may be actually connected via the root system as a consequence. Root grafting between initially separate individuals can produce the same effect.

In herbaceous plants, however, the root connections between developing root buds are likely to decay and individual plants will lose contact with each other.

Fig. 178. *Rubus idaeus*
Shoot apical meristems emerging from within root tissue (endogenous development).

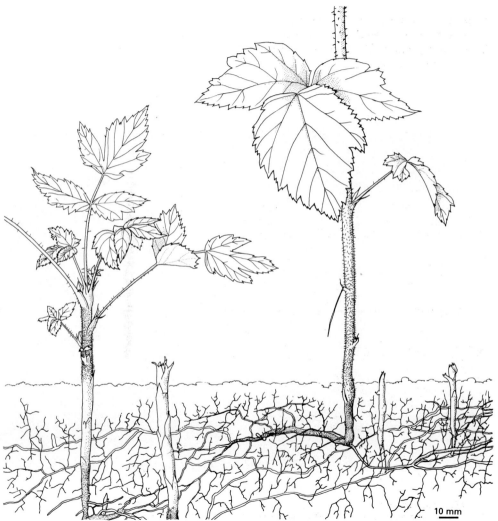

Fig. 179. *Rubus idaeus*. Various stages of shoot development from root buds.

10 mm

A young grass seedling consists of a very short stem axis virtually lacking internodes such that the nodes bearing leaves are very close together. Each leaf is attached at its node around all or most of the stem circumference and thus forms a tube, the leaf sheath, which may be open on one side, and which surrounds the next youngest leaf. The first leaf of the seedling is represented by a reduced absorptive organ contained within the grass fruit termed the scutellum. The second leaf emerges above ground on germination and forms a simple green tube, the coleoptile with a hole in its apex (**163b** cf. **164**). Subsequent leaves each develop to a greater or lesser extent in two stages. The part of the leaf which is formed first and therefore becomes the distal end is usually flat and termed the lamina. It increases in length due to cell division at its proximal (**4**) end (an intercalary meristem **18**). The lower proximal end of the leaf forms the leaf sheath, is the more or less tubular part, and is distally flattened in some species; it also has an intercalary meristem at the lowest proximal point. The lamina can bend backwards relative to the sheath at a sometimes distinctive zone of cells, the lamina joint. Also found at the junction of lamina (or blade) and sheath is a flange of tissue, the ligule (**181h**), which sometimes may be replaced by a fringe of hairs (**181b, c**) or is absent. Outgrowths at the side of the lamina joint region are termed auricles (**181h**). Some grass leaves (and especially bamboos **192**) have rather wide short laminas with a distinct petiole connecting sheath to lamina. This petiole is not homologous with the petiole of dicotyledon leaves (**20**). Leaves on a

grass stem are located in two rows on opposite sides (distichous phyllotaxis **219c**) and the young leaves will be folded together in an imbricate and usually equitant manner (vernation **39g**) depending upon the degree of folding along the mid line. As a young grass plant develops, longer internodes will be formed either during the

production of a vertical inflorescence axis or between successive leaves of a rhizome (**131f**) or stolon (**133d**). The nodes on the vertical axis, the culm, often appear swollen and form adjustable joints. It is actually the very proximal end of the leaf sheath which forms the swollen portion, i.e. a leaf pulvinus (**46**). Swollen stem tubers

Fig. 180. *Dactylis glomerata* var. *hispanica*
A prostrate form of a tussock forming grass. A tiller (side shoot)
is developing from the axil of every leaf on the main axis.

(181e, f) are formed in some grasses. The leaf sheath attached at a node will often be tightly folded or rolled around the next internode and also encircled by sheaths attached at lower nodes. To see which leaf belongs to which node it is thus necessary to pull these structures away from each other. This will also allow the identification of the buds in the leaf sheath axils, and the point of insertion of side shoots (tillers 182) to be determined. All the roots on a grass plant (except the very first, primary, seminal root which is protected by a dome of tissue termed the coleorhiza) are adventitious (98), that is they are formed from root primordia developing in the stem usually at leaf nodes. Adventitious root primordia present in the embryo before germination, usually located at the nodes of the coleoptile and first foliage leaf, are termed lateral seminal roots. Subsequent roots are referred to as nodal. The majority of grasses branch repeatedly, lateral daughter branches (tillers) usually having the same general morphology as their parent (182).

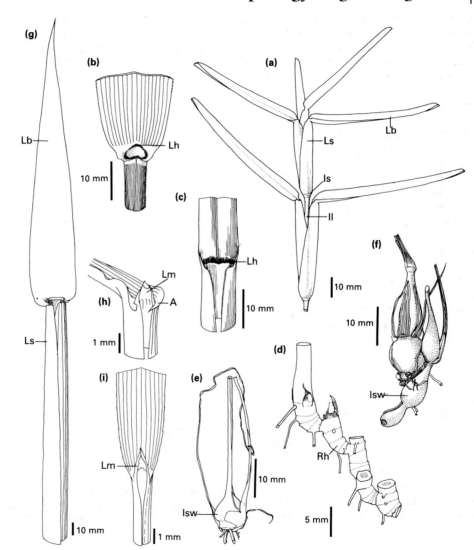

Fig. 181. a) *Stenotaphrum secundatum*, vegetative shoot with alternating long and short internodes; b) *Phragmites communis*, leaf blade/sheath junction; c) *Cortaderia argentea*, leaf blade/sheath junction; d) *Stenotaphrum secundatum*, rhizome (130); e) *Panicum bulbosum*, base of swollen tiller; f) *Arrhenatherum elatius* var. *bulbosum*, series of swollen internodes (cf. pseudobulb 198); g) *Arundo donax*, single leaf; h) *Lolium perenne*, leaf blade/sheath junction; i) *Poa annua*, leaf blade/sheath junction. A: auricle. Il: long internode. Is: short internode. Isw: swollen internode. Lb: leaf blade. Lh: hairy ligule. Lm: membranous ligule. Ls: leaf sheath. Rh: rhizome.

It is customary to refer to the first shoot of a grass plant, i.e. the axis developing from the epicotyl (162) as the parent shoot. All subsequent shoots must develop from axillary buds and are termed tillers. The first leaf (prophyll 66) on a tiller is usually much smaller than later formed leaves and may not be constructed of distinct lamina and blade. It is in an adaxial position (4). A tiller may develop in such a manner that it closely resembles the parent shoot, growing vertically and bearing a terminal inflorescence. Its vertical growth will cause it to extend up trapped between the parent shoot axis and the sheath of its subtending leaf (**183a**). This type of development is termed intravaginal. Each leaf at the base of the parent shoot, including the coleoptile, can subtend such a tiller. Subsequently, buds in the axils of leaves at the base of each tiller may themselves develop into tillers. A compact grass plant will thus be formed. Adventitious roots will develop from lower nodes of both parent shoot and tillers. Each tiller will thus have its own set of leaves, roots, and daughter tillers. Compact intravaginal growth gives a 'caespitose' ('clumping', 'bunch', 'tussock', 'tufted') habit. Conversely a tiller may grow out sideways away from the parent shoot and therefore be more or less at right angles to it. This results in the tiller breaking through the base of its subtending leaf sheath and is referred to as extravaginal tillering (**183b**). The tiller so formed will usually be a procumbent shoot lying on the ground or on surrounding vegetation (cf. **182**), or grow strongly and horizontally away from the parent above ground (a stolon **133d**) or

beneath ground (a rhizome **131f**, as is typical for the bamboos **195**). The leaves on such horizontal axes are often cataphylls (scale leaves 64) especially if the tiller is underground. Eventually the tip of the extravaginal tiller will turn to grow erect, horizontal growth being continued by a daughter tiller (i.e. sympodial growth 250). Nodal roots are likely to form on horizontal tillers.

Leaf arrangement in a grass plant is distichous (**219c**). A plan diagram of a tiller system should therefore appear as in Fig. **183d**, with all leaves formed in one plane and a fan-shaped tussock developed. This does not always occur, however, as each tiller bud is displaced (230) somewhat round the node at which it is attached, i.e. it is not in line with the mid-point of its subtending leaf (**183e**). This has a corresponding effect on the shape of the grass tussock. Developmental studies of tillering in grasses (including cereals) are often aided by some system of ordering (284) for the tillering sequences. The leaves on the parent shoot are recorded as C (for 'coleoptile'), L1, L2, etc. The tiller in the axil of the coleoptile is designated TC, that in the axil of the first foliage leaf T1, and so on. If T1 itself bears tillers, the first will be in the axil of the prophyll of tiller 1 and can be identified as T1.PT. The next tiller will be in the axil of the first foliage leaf of tiller 1, and is referred to as T1.L1T. Figure **183c** illustrates how such a labelling system can be built up should it be necessary to refer to any individual tiller or leaf accurately. (A similar system can of course be designed for any type of plant.) Detailed aspects of grass (specifically cereal) morphology are given in Kirby (1986).

Fig. 182. *Arundo donax*
A stout grass which is unusual in that vegetative shoots develop from the aerial stem. Each tiller is breaking through its subtending leaf sheath, a situation normally found in rhizomatous or stoloniferous grasses (cf. Fig. **183b**). This specimen is variegated.

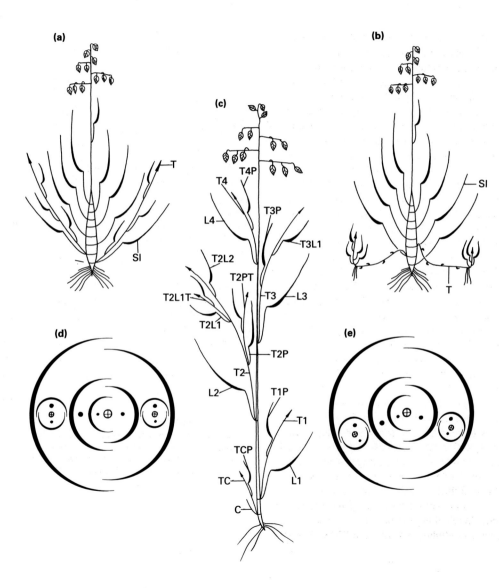

Fig. 183. a) Intravaginal tillering; b) extravaginal tillering; c) labelling system; d) plan diagram hypothetical distichous arrangement of leaves with buds in line with mid-veins; e) typical plan diagram showing displacement of buds around axis away from mid-line. Sl: subtending leaf. T: tiller.

Fig. 184. *Aegilops ovata*
The rounded outer glume of each spikelet bears long terminal awns.

Inflorescences are categorized on the basis of the arrangement of flowers and the pattern of branching (140). Flowers in grasses and bamboos, however, are aggregated into groups enclosed between a pair of scale leaves (glumes). Each such package is termed a spikelet (186), and it is the arrangement of spikelets that is used to describe the grass inflorescence rather than the arrangement of individual flowers. Similarly if a spikelet is borne on a stalk, the stalk is referred to as a pedicel although this term is normally applied to the stalk of an individual flower. The stem with elongated internodes supporting an inflorescence is termed a culm. The culm also bears vegetative branches in bamboos and other stout grasses (**182**). The main axis of the inflorescence is termed the r(h)achis. This may be variously branched and the branches relatively long or short or both, the general form being a raceme (panicle if repeatedly branched **141g, 185d, e, g, h, i**). The nodes at which individual branches occur can be grouped very close together at intervals along the rachis, branches appearing to be attached at one point. The spikelets themselves may be carried on long or short pedicels, or be sessile forming a spike (**185a, b, c**). If a number of spikes are all apparently attached at one point, a digitate arrangement results. The inflorescence of barley (*Hordeum* spp.) consists of an axis bearing two rows (one on each side) of spikelets grouped in threes on extremely short pedicels (**189j**). In wheat (*Triticum* spp.) two rows of solitary sessile spikelets are present (**188c**). In such a spike, the sessile spikelet may be sunken into the rachis or the two rows may appear side by side rather than front to back due to displacement during development. Each spikelet forms from one bud but its axillant leaf (bract 62) is usually missing although some evidence of a ridge of tissue may be visible. The 'collar' at the base of a wheat inflorescence represents such a subtending bract. The extreme distal end of the rachis or side branch may terminate in a fully formed spikelet, or a partly complete sterile spikelet, or may terminate blindly in a non-meristematic point. When the fruits are mature a grass inflorescence usually falls apart. Points of articulation (breakage) vary amongst species. Single spikelets with or without their glumes (186) may be shed, or groups of spikelets or whole spikes or whole inflorescences may fall. Some spikes break into segments each bearing just one spikelet.

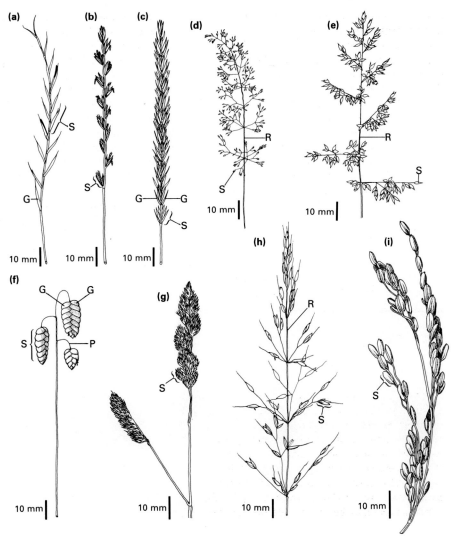

Fig. 185. Example of variation in grass inflorescence structure. a–c) spikes, d–i) panicles: a) *Nardus stricta*, b) *Lolium perenne*, c) *Agropyron* (*Elymus*) *repens*, d) *Agrostis tenuis*, e) *Holcus lanatus*, f) *Briza maxima*, g) *Dactylis glomerata*, h) *Arrhenatherum elatius*, i) *Oryza sativa*. C: culm. G: glume. P: pedicel. R: rhachis. S: spikelet.

An individual grass flower contains a superior ovary (146) with three (two or one) styles, three (two or one) stamens (six in bamboos and a few other grasses), and two (occasionally three or none) small structures, lodicules, representing perianth segments (146). Unisexual flowers also occur in some species. Flowers are borne in groups in numbers characteristic for a species (from one to many), along a short axis, the rachilla. Each flower is in the axil of a bract, termed the lemma, and the stalk of each flower bears a bracteole, the palea (**187j**). The lemma usually envelops the palea, the two protecting the flower inside which is only visible when the lodicules swell up forcing the lemma and palea apart and exposing the anthers and the style (**186**). Grasses are almost all wind pollinated (cf. **192b**). The whole structure, lemma, palea, and flower, is referred to as a floret (**187j**). The rachilla has at its base two extra bracts which are sterile, i.e. they do not subtend florets. The most proximal bract, which is the prophyll of the rachilla but which is not necessarily in an adaxial position, is called the lower glume. The second sterile bract is the upper glume. This pair of glumes will to a varying extent envelop a characteristic number of florets. The whole unit, glumes plus florets, is termed the spikelet (**187k**) and is a consistent feature of the grasses. Grass inflorescences are described according to the arrangement of spikelets rather than that of individual flowers (184). When studying a grass inflorescence the first step will always be the identification of spikelet units by locating pairs of glumes at their bases. An individual spikelet may contain different types of floret (*Sorghum* **191a**) or the inflorescence may contain different types of spikelet, for example fertile and sterile (*Cynosurus* **187e, f**). In *Setaria* and *Pennisetum* (**191b**) the terminal spikelets are missing and only their stalks are present as bristles (referred to incorrectly as an involucre 146). The inflorescence of *Coix* is represented by a limited number of branches each terminating in a hard glossy bead-like structure. The 'bead' represents the strengthened base of the leaf subtending a short rachis. The rachis bears one female spikelet, which remains encased by the bead, and a series of several male spikelets borne on an axis that protrudes out of the bead.

Glumes, lemmas, and paleas are typically membraneous cataphylls (64) and vary in shape and complexity; one or more of these structures may be absent from a spikelet (**185b**). They may be rounded or keeled, that is folded along the mid line, i.e. conduplicate (**37j**), and may bear an awn which is an extension of the mid vein. The awn is located at the tip of the lemma (**189h**) or glume (**189d**) or the vein departs from some point on the dorsal side. The base of a dorsal awn is often much twisted and the awn itself kinked above the twisted portion. Such a geniculate awn (**187d**) responds to drying or wetting by rotating and levering the associated fruit amongst vegetation or soil particles. If several veins end in awns they may be spread apart (**184, 187h**) or be twisted together for part of their length. Hairs are often present in and amongst spikelets and can appear to be the most conspicuous part of the inflorescence. Florets are occasionally replaced by small tillers complete with adventitious roots (false vivipary 176).

Fig. 186. *Arrhenatherum elatius*
Anthers and stigmas protruding from the fertile hermaphrodite spikelets. One floret in each such spikelet has a lemma bearing an awn, several of which are visible here.

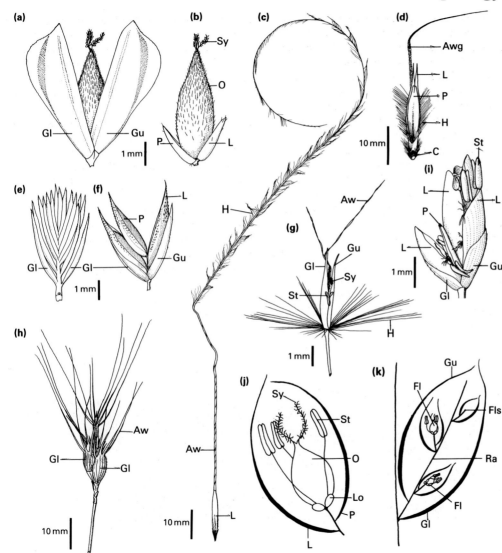

Fig. 187. a) *Phalaris canariensis*, single spikelet; b) *Phalaris canariensis*, single floret; c) *Stipa pennata*, single floret (awned lemma); d) *Avena* sp., single floret (geniculate awn on back of lemma); e) *Cynosurus cristatus*, sterile spikelet; f) *Cynosurus cristatus*, fertile spikelet; g) *Miscanthus*, sp., single spikelet; h) *Aegilops ovata*, group of spikelets; i) *Poa annua*, single spikelet; j) diagram of single floret; k) diagram of single spikelet. Aw: awn. Awg: geniculate awn. C: cicitrix. Fl: floret. Fls: sterile floret. Gl: lower glume. Gu: upper glume. H: hair. L: lemma, Lo: lodicule. O: ovary. P: palea. Ra: rachilla. St: stamen. Sy: style.

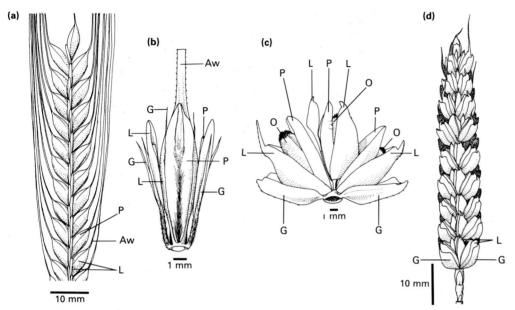

Fig. 188. a) *Hordeum vulgare* var. *distichum*, part of an inflorescence spike; b) *Hordeum vulgare*, abaxial view cluster of three spikelets (one fertile, two sterile); c) *Triticum aestivum*, single spikelet; d) *Triticum aestivum*, inflorescence spike. Aw: awn. G: glume. L: lemma. O: ovary. P: palea.

A cereal is a cultivated species of grass, its morphology invariably modified by selection. Nevertheless all the features of grass vegetative growth (180, 182), inflorescence (184), and spikelet construction (186) apply to the cereal plant. The inflorescence of wheat (*Triticum* spp.) is a spike (**188d**) in which one spikelet only (**188c**) is present at each node as can be detected by the presence of two glumes at each node. Between each pair of glumes is a short axis (rachilla) bearing a limited number of hermaphrodite florets. Two, three, or occasionally four of the lower florets will be fertile. Spikelets alternate on either side of the inflorescence axis (rachis) and there is usually a terminal spikelet present. Differences in appearance between species and varieties of wheat depend upon the number of fertile florets per spikelet, the compactness of the inflorescence spike, and the extent of awn development on the lemmas, and occasionally on the glumes also. The wheat grain (a caryopsis fruit **157b**) falls out from between its lemma and palea when threshed. The inflorescence of rye (*Secale* sp.) is very similar to that of wheat but habitually each spikelet contains two fertile florets and the remains of a sterile floret (**189a**). An inflorescence of barley (*Hordeum* sp.) may resemble that of wheat. It is a spike in which spikelets occur in groups of three on a very short axis at each node (**188a**). This axis itself ends without a terminal spikelet, all spikelets being lateral. Nevertheless the three spikelets are arranged such that one appears to be central, flanked by the other two. The two lateral spikelets may be sterile (**188b**). Each spikelet contains only one floret. Thus at each node, on alternating sides of the rachis, there are to be found three pairs of glumes, often very small, particularly if associated with lateral sterile spikelets, and three sets of lemma plus palea. The lemmas invariably bear long awns or occasionally elaborate hooded distal ends (**189e[1]**). The hood of a hooded barley may incorporate an extra epiphyllous (74) spikelet. If all three spikelets at a node are fertile the inflorescence spike appears to consist of six (three each side) rows of spikelets (**189f, f[1]**). If the lateral spikelets are sterile, the inflorescence appears to be composed of two rows of spikelets only (a central one on each side) (**189h, h[1]**), although the glumes of the lateral spikelets will be found in place. An apparently four-rowed barley results if the lateral fertile spikelets of one side interdigitate with the lateral spikelets of the other (**189g, g[1]**). Other commonly cultivated cereals are described in the following section. (Continued on page 190.)

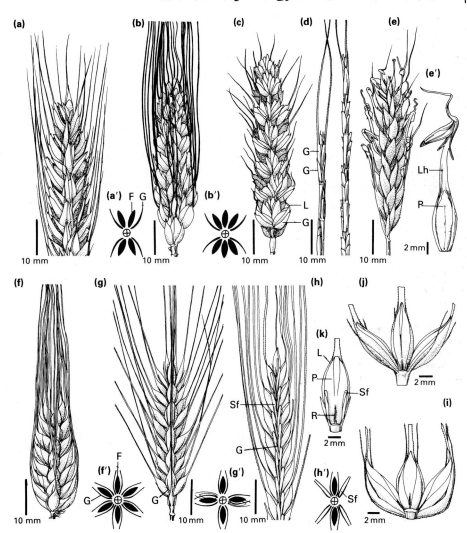

Fig. 189. a) *Secale cereale*, part of inflorescence spike; a[1]) diagram of spikelet layout; b) *Triticum durum*, inflorescence spike, b[1]) spikelet layout; c) *Triticum* sp. (Nepalese); d) *Aegilops speltoides*, inflorescence spike; e) *Hordeum* sp. (hooded), e[1]) *Hordeum* sp. (hooded) single floret; f) *Hordeum vulgare* var. *hexastichum*, f[1]) spikelet layout; g) *Hordeum vulgare* var. *tetrastichum*, g[1]) spikelet layout; h) *Hordeum vulgare* var. *distichum*, h[1]) spikelet layout; i) as f), cluster of three spikelets; j) as g), cluster of three spikelets; k) as h), cluster of three spikelets. F: floret. G: glume. L: lemma. Lh: hooded lemma. P: palea. R: end of axis bearing spikelets. Sf: sterile spikelet.

(a)

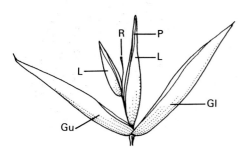

(b)

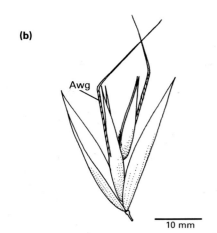

10 mm

Fig. 190. a) *Avena* sp., single spikelet (lemmas awnless); b) *Avena sativa*, single spikelet. Awg: geniculate awn. Gl: lower glume. Gu: upper glume. L: lemma. P: palea. R: rachilla.

Avena sativa (Oat)

The inflorescence of oat forms a loose panicle (**141g, 185h**) with crowded nodes such that pseudowhorls of branches may be present. Each ultimate branch ends in a conspicuous spikelet (**186**) which demonstrates spikelet construction very clearly. The glumes are especially large and protect a rachilla bearing up to 7 florets of which the lower 1, 2, or 3 may be fertile. Paleas are relatively inconspicuous, but the lemma may bear an awn, often geniculate (**190b**). The base of the lemma is variously swollen (the callus) and bears hairs. The floret often disarticulates at this point possibly leaving a prominent scar, the cicatrix (**187d**).

Oryza sativa (Rice)

The inflorescence is a panicle partly enclosed by the most distal leaf of the culm (**185i**). Each branch of the panicle terminates with a spikelet containing one fertile (rarely more) floret. Six stamens are present (most grasses have 1, 2, or 3; bamboos also have 6). The glumes are small (**191c**), the lemmas variably awned. The spikelet stalks (pedicels **184**) break below the spikelet.

Zea mays (Maize, Corn)

Maize has two distinct inflorescence forms: totally female inflorescences ('ear', 'cob') borne in the axils of leaves on the culm, and a terminal male inflorescence ('tassel'). The latter is a panicle with pairs of similar spikelets, one sessile and one on a short pedicel. Each spikelet consists of a pair of glumes surrounding two male florets. The female inflorescence is a spike contained within a

number of large scale leaves (the husk) borne on the proximal end of its rachis. Female spikelets occur in pairs and each spikelet contains one sterile and one fertile floret. Glumes, lemmas, and paleas are all shorter than the large ovary. Styles (silk) are very long and emerge from the distal ends of the ensheathing husk.

Sorghum bicolor (Sorghum, Great Millet) ('Millet' is a general term covering many distinct cereal grain species with numerous common names)

The inflorescence is a panicle with spikelets borne in pairs, one hermaphrodite and sessile, one male or sterile and borne on a short pedicel (**191a**). The hermaphrodite spikelet is considerably larger than the male spikelet and contains two florets. The lower floret is sterile and lacks a palea, the upper one has a lemma but again the palea can be absent. The stalked male spikelet likewise contains a lower sterile floret represented by a lemma only, and the upper male floret also without palea. In some forms this spikelet consists of a pair of glumes only.

Panicum miliaceum (Common Millet, Proso Millet)

The inflorescence is a panicle, with solitary spikelets. The upper glume of each spikelet is longer than the lower and envelops a lower sterile floret and an upper fertile floret (**191f**).

Pennisetum typhoides (Bulrush Millet, Pearl Millet)

The inflorescence is a dense panicle or loose branched spike. The spikelets occur in pairs aggregated into dense groups. Proximal to each pair are borne numerous pedicels lacking terminal spikelets and forming a mass of bristles (termed an 'involucre', see comment for *Setaria* 186) varying greatly in length in different varieties (**191b**). Each spikelet contains one male floret and one hermaphrodite floret.

Eleusine coracana (Finger Millet, African Millet) The inflorescence consists of a radiating cluster of spikes located on the top of the culm. Each spike bears two rows of spikelets on the outer side of its rachis. The rows overlap each other to some extent. Each spikelet, identified by its basal pair of glumes, contains up to a dozen florets which are borne left and right on the spikelet rachilla (**191d**). The florets are hermaphrodite and lemmas and paleas conspicuous.

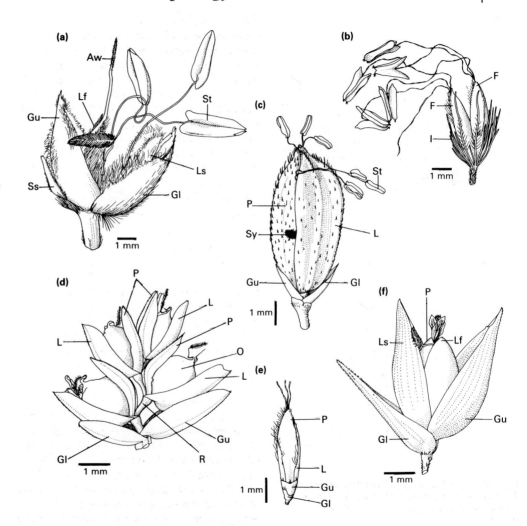

Fig. 191. a) *Sorghum bicolor*, spikelet pair; b) *Pennisetum typhoides*, spikelet pair; c) *Oryza sativa*, single spikelet; d) *Eleusine coracana*, single spikelet; e) *Pennisetum typhoides*, single spikelet anthers removed; f) *Panicum miliaceum*, single spikelet. Aw: awn. F: floret. Gl: lower glume. Gu: upper glume. I: involucre. L: lemma. Lf: lemma of fertile floret. Ls: lemma of sterile floret. O: ovary. P: palea. R: rachilla. Ss: sterile spikelet. St: stamen. Sy: style.

Fig. 192a. *Bambusa arundinacea*
The branching inflorescence of a dying plant (cf. Fig. **194**) (the leaves in the foreground are of a palm).

Fig. 192b. *Piresia* sp.
An entire plant: one of the smallest bamboos. Leaf litter has been removed to expose a plagiotropic (246) underground inflorescence which bears scale leaves and, in this specimen, two distal spikelets. Pollination is probably performed by ants.

Bamboos are grasses (family Gramineae, tribe Bambuseae) and can usually be recognized by a combination of woodiness and persistence of both culm (180) and rhizome (194), by the vegetative branching of the culm, by a short petiole between sheath and blade of vegetative leaves (**193b**), and by spikelets containing more component parts than other grasses (i.e. possibly >2 glumes, or sterile lemmas, >2 lodicules, >3 stamens, >2 styles **193f, g**). The bamboo culm consists of a series of more or less elongated internodes separating distichously (**219c**) arranged scale leaves at the nodes. The scale leaf represents the sheath only of a complete leaf and may have a small portion of lamina at its distal end (**193d**) together with a ligule and auricles depending upon the species and its position in the heteroblastic sequence (28). Each scale leaf subtends a bud (**193e**). Such vegetative buds usually bear their own buds and develop into an elaborate condensed branching system (**193c, 239c**). Some species also possess true accessory buds (236). Branch complexes are persistent and continue to branch on a seasonal basis. Scale leaves on the culm and prophylls within the lateral branching complexes fall easily but are then represented by prominent scars (**193c**). Dormant buds are frequently sunken into a more or less prominent groove. The lateral branches may take the form of stem spines (124), or slender vegetative branches bearing foliage leaves or inflorescences in flowering individuals (**192a, 193a**). Bamboo infloresences, borne laterally on the culm branches, are mostly panicles (**141g**) but many incorporate sub-units of sessile spikelets. Lower spikelets in a group may be replaced by a reserve bud (pseudospikelet; McClure 1966) allowing additional inflorescence branches to occur (indeterminate inflorescence; McClure 1966); in the absence of these extra buds the inflorescence will be determinate (McClure 1966). Bamboos are wind pollinated; an exception is *Piresia* (**192b**).

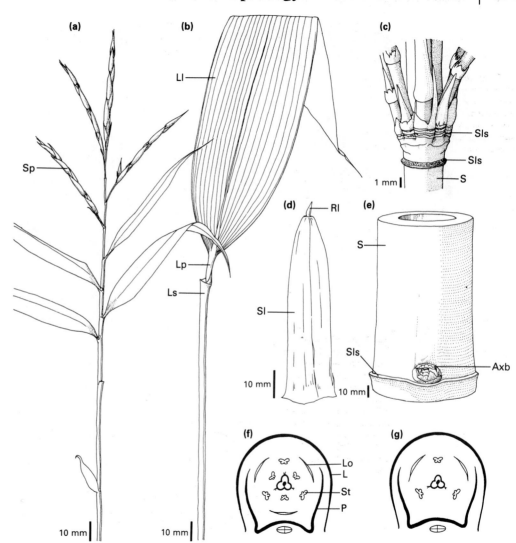

Fig. 193. a) *Arundinaria* sp., flowering tiller; b) *Sasa palmata*, single foliage leaf; c) *Sinarundinaria* sp., condensed branching on aerial shoot (see **239c**); d) *Sasa palmata*, single scale leaf; e) *Bambusa arundinacea*, section of aerial shoot; f) floral diagram bamboo spikelet; g) floral diagram grass spikelet. Axb: axillary bud. L: lemma. Ll: leaf lamina. Lo: lodicule. Lp: leaf petiole. Ls: leaf sheath. P: palea. Rl: rudimentary lamina. S: stem. Sl: scale leaf. Sls: scale leaf scar. Sp: spikelet. St: stamen.

Fig. 194a, b. *Bambusa arundinacea*
a) Excavated rhizome system from dying clump. One aerial culm is present at the top of the picture. Model of McClure (**295c**).

b) Close-up of upper surface of rhizome segment showing rows of dead adventitious roots (98) alternating with scale leaf scars in which vein endings can be seen.

Bamboos develop extensive and persistent woody underground branching rhizome systems. The rhizome branch bears scale leaves (64) only, and adventitious roots (98) are produced extensively at the nodes (**194b**). Two basic rhizome types are recognized by McClure (1966). (i) Pachymorph (cf. pachycaul, 130)—short and fat rhizome branches, usually solid and terminating distally in a vertical culm (**194a**). Buds on these rhizomes always give rise to other rhizome branches (**195a**). (ii) Leptomorph—long and thin, usually hollow and extending more or less indefinitely underground, i.e. rarely turning erect to form a terminal culm. Buds on these rhizome branches usually become aerial culms or occasionally additional underground leptomorph rhizome branches (**195d**). The proximal end of any new rhizome branch or lateral culm is always relatively thin and referred to as the rhizome, or culm, neck (**195b**). The neck is often orientated somewhat downwards (**195a**) especially in seedlings (see establishment growth 168) and usually has no buds in the axils of its scale leaves and no adventitious roots. The neck of a pachymorph rhizome may be short (**195a**) (or long **195b**); that of a leptomorph rhizome is always short (**195d**). The junction of culm neck and culm may be extended by a series of short internodes termed by McClure (1966) a metamorph axis type 1 (**195c**). These only occur on laterally borne culms. The distal ends of pachymorph rhizomes and of leptomorph rhizomes that terminate in culms, may be extended by a series of long internodes termed by McClure (1966) metamorph axis type 2 (**195f**).

Combinations of these distinctive features are to be found in different bamboo species (**195e, g**). The non-woody and non-persistent underground parts of other members of the grass family often show similar morphologies to those found in the bamboos (**181d**), and both form comparable branching patterns to the rhizome systems of gingers (**311**).

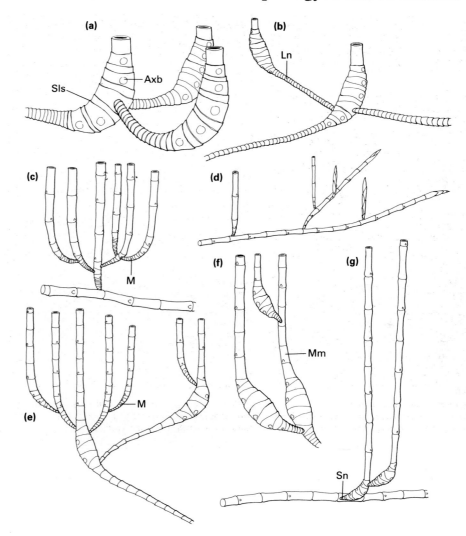

Fig. 195. Bamboo rhizome types. Redrawn from McClure (1966). a) pachymorph; b) pachymorph, long neck; c) metamorph axis type 1; d) leptomorph; e) pachymorph, long neck and metamorph type 1; f) metamorph axis type 2; g) leptomorph and pachymorph, short neck. Axb: axillary bud. Ln: long neck. M: metamorph type 1. Mm: metamorph type 2. Sls: scale leaf scar. Sn: short neck.

Members of the sedge family (Cyperaceae) show a considerable range of vegetative and reproductive morphology and many superficially resemble grasses (180). Vegetative leaves have a sheath (forming an entire cylinder around the stem) and a narrow blade with a ligule at the junction of the two. The stem is usually solid and the leaves are borne on it in three rows (tristichous **219e**, grasses are distichous **219c**). The aerial shoots of sedges invariably represent the distal ends of the underground sympodial rhizome segments (**269d**). These rhizomes are of many types, pachymorph and leptomorph (see bamboo terminology 194) or may incorporate stem tubers (138) at the base of the aerial shoot. The tristichous leaf arrangement frequently governs the directional spread of successive rhizome branches in the sympodial sequence (**197c, c¹**). Also, successive sympodial units are frequently adnate for part of their length giving a misleading monopodial appearance particularly in leptomorph species (**235a**). The inflorescence is of variable construction again showing much the same overall range of types as those of the grasses (184). Individual flowers, hermaphrodite, male, or female, are found in characteristic positions within the ultimate units of the inflorescence (termed spikelets as in grasses 186) and lack perianth segments or else these are represented by bristles or scales. Flowers themselves are subtended by scale leaves ('glumes'). Interpretation of the detailed arrangement of parts is usually facilitated by the presence of prophylls in the usual adaxial position (66), but arrangements are not as

consistent as in that of the grass spikelet (186). In the genus *Carex* and others the prophyll subtending the female flower is a large flask-shaped structure, the utricle which surrounds the flower. Such a female flower is lateral on a rachilla which may be a conspicuous feature within the utricle. The units of the inflorescence within the Cyperaceae are discussed at length by Eiten (1976); a grossly simplified selection of the range to be found is shown in Fig. **197d–i**.

Fig. 196. *Bulbostylis vestita*
The vertical stem is protected from natural fire in its savanna habitat by the mass of persistent leaf sheaths. Model of Corner (**291d**).

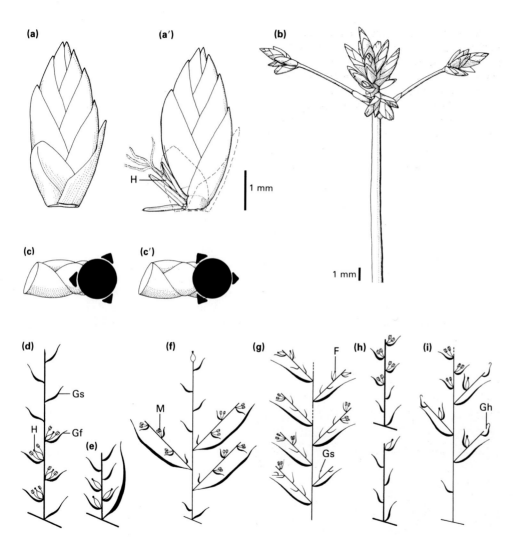

Fig. 197. a) *Cyperus alternifolius*, single spikelet; a¹) ditto, with glumes removed; b) *Cyperus alternifolius*, inflorescence; c, c¹) *Eriophorum*, alternative rhizome bud locations at base of aerial shoot; d–i) redrawn from Eiten (1976), selection of sedge floral types. F: female flower. Gf: fertile glume. Gh: hooked glume. Gs: sterile glume. H: hermaphrodite flower. M: male flower.

Fig. 198. *Campylocentrum pachyrhizum*
The shoot system of this orchid is at the centre of the picture together with dead inflorescence axes. The bulk of the plant consists of green flattened adventitious roots (98).

Orchid species (Orchidaceae) have distinctive and usually elaborate flowers (200). Vegetatively they show a range of forms outlined here as an example of constructional variation in one distinctive taxonomic group (see also 253). The majority of orchids have either a sympodial or less frequently monopodial rhizome although as the plant may be epiphytic this stem system will not be below ground (170). Monopodial orchids have lateral inflorescences (253b, 199b), sympodial orchids have lateral (253a, d, 199d) or terminal (253c, 199e) inflorescences. A distinctive feature of many orchids is the pseudobulb (199d–f). This represents a swollen segment of stem of one or more internodes and is thus morphologically equivalent to a corm (137d). The location of a pseudobulb within the rhizome system of a particular orchid is usually very precise as may be the number of leaves it bears, and shows a range of permutations (199, 253). Orchid leaves are variable in shape in different species and scale leaves are often present. In addition to storage pseudobulbs, orchids may possess variously swollen roots (root tuber 100). In some instances these incorporate stem tissue with a shoot apex and are equivalent to the droppers (175f, g) of other plants. A second feature of many orchid roots is an extensive water absorbing covering, the velamen (106). Such roots may also be photosynthetic; an extreme case, *Campylocentrum*, is shown in Fig. 198.

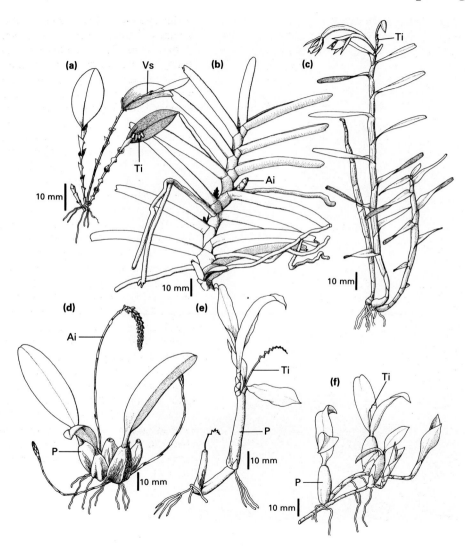

Fig. 199. Examples of growth form variations of Orchids. a–c) without pseudobulbs, d–f) with pseudobulbs. See also **253**. a) *Restrepia ciliata*, b) *Acampe* sp., c) *Epidendrum* sp., d) *Bulbophyllum* sp., e) *Pholidota* sp., f) *Coelogyne fimbriata*. Ai: axillary inflorescence. P: pseudobulb. Ti: terminal inflorescence. Vs: vegetative shoot.

The orchid flower exhibits a number of features that collectively define the family, although they are also found in part in other groups. Dressler (1981) lists seven characteristics:

(1) Stamens on one side of flower (usually one active);
(2) Stamens adnate to pistil (=column) (**201d¹**)
(3) Petal opposite stamen elaborate (cf. **201b**) (=labellum or lip);
(4) Part of stigma represents pollination apparatus (=rostellum) (**201d¹**).
(5) Pollen massed into pollinia (**201a¹**);
(6) Flower stalk (pedicel) often twists (resupination) (**201e**);
(7) Extremely small seeds (**159h**).

An orchid flower is composed of six perianth segments in two whorls of three. The adaxial petal of the inner whorl is the elaborated labellum. Twisting of the pedicel (resupination) turns the flower through 180° in most cases bringing the labellum into an apparently abaxial position. One (sometimes two) stamen unites with the style to form a column. The upper part of the column bears the elaborated anther (clinandrium) and elaborated stigma (rostellum). Flowers may be solitary, or may be aggregated into inflorescences. Arrangement is usually racemose (**141b**), occasionally cymose (**141o**), and rarely leaf opposed (**230**).

Fig. 200. *Paphiopedilum venustum*

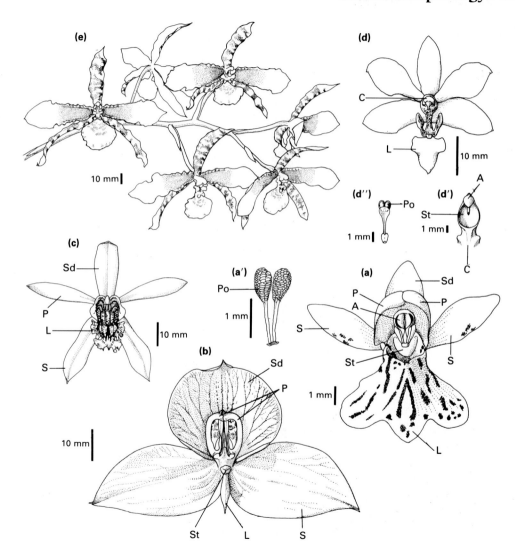

Fig. 201. a) *Dactylorhiza fuchsii*, flower; a¹) *Dactylorhiza fuchsii*, pollinia; b) *Disa* 'Diores'; c) *Coelogyne* sp., flower; d) *Doritis pulcherrima*, flower; d¹) *Doritis pulcherrima*, column; d¹¹) *Doritis pulcherrima*, pollinia; e) *Rossioglossum grande*, inflorescence showing resupination. A: anther (clinandrium) associated with rostellum. C: column. L: labellum (lip). P: lateral petal. Po: pollinium. S: lateral sepal. Sd: dorsal sepal. St: stigmatic surface.

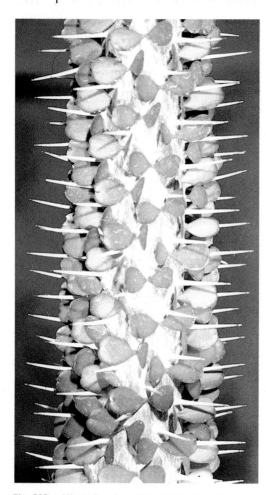

Fig. 202a. *Alluaudia adscendens* (Didiereaceae)
Each pair of leaves with associated stem spine (124) has developed from the bud in the axil of a leaf now shed.

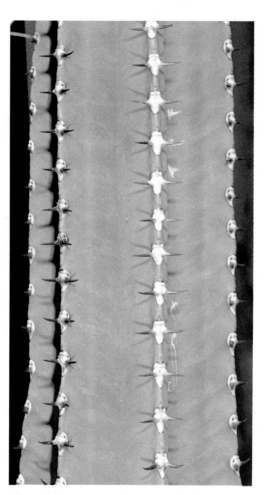

Fig. 202b. *Euphorbia ammati* (Euphorbiaceae)
Each pair of spines represents a pair of stipules (52). A minute bud is present in the axil of each leaf scar.

The spines of cacti (Cactaceae) represent modified leaves. In *Pereskia* species, normal bifacial leaves are present with the leaves of their axillary buds developing as spines. Two spines only per bud in some species (**203b**) represent modified prophylls (66). In the majority of cacti, the green stem is either flattened (and thus can be termed a phylloclade 126, **203a**, **294a**) or is swollen with conspicuous protuberances ('tubercles', 'mammillae', leaf cushions **203g**). These tubercles may merge into vertical ridges (**203c**). In flattened *Opuntia* spp., small and temporary leaves can be seen on newly developing stems (**203a**). Each leaf subtends an axillary bud, the leaves of which are again represented by groups of spines. Each such group of spines is termed an areole (cf. 34). Some spines (glochids) are barbed and are easily detached. In species with tubercles (mammillae) an areole is usually found on the distal end of this structure having originated in the axil of a leaf whose tissue is now incorporated in the tubercle. Such meristematic activity (leading to the development of combined tissue) is an example of adnation (234) and in this case results in an axillary bud situated on its subtending leaf (epiphylly 74). Some species of the genus *Mammillaria* exhibit another form of meristem reorganization leading to the symmetrical division of the apex, i.e. true dichotomy (258). The apical meristem of the areole may die, remain dormant, continue to produce more leaves as spines, or develop into another vegetative shoot or a flower. In some species two buds (i.e. accessory buds 236) are associated with each leaf site, one developing into the areole on the tubercle, the other having the potential to become either a vegetative shoot or a flower or occasionally a second areole (**203d**). This second bud will be found somewhere on the adaxial side of the tubercle. All the leaves of a bud forming an areole do not necessarily develop into similarly sized spines. The spines on the abaxial side of the areole are usually the largest. Hairs (trichomes 80) may be found amongst the spines. Members of the related family Didiereaceae similarly have the leaves of axillary buds modified in the form of spines (**202a**). Some members of the Euphorbiaceae (**203f, h**) and Asclepiadaceae (*Stapelia* and *Ceropegia* **203e**) resemble cacti with swollen, flattened, ridged, or tubercled stems. Spines present in the Euphorbiaceae are either present in pairs and then represent modified stipules (**202b**), or are solitary in the axil of a leaf or leaf scar and then represent modified inflorescence axes (144) or persistent leaf bases (40). Cactus-like members of the Asclepiadaceae bear dormant vegetative or reproductive buds in the axils of leaf spines.

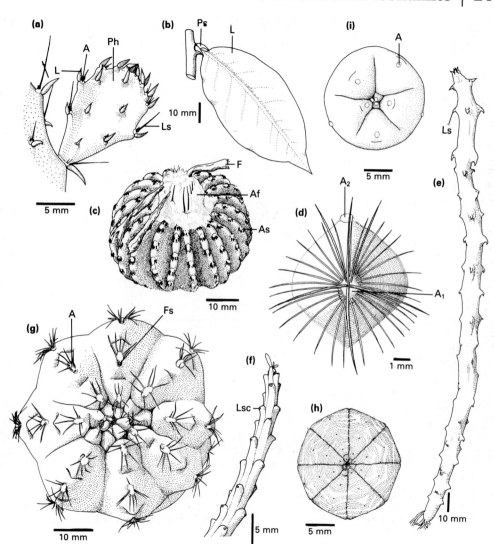

Fig. 203. a) *Opuntia* sp.; b) *Pereskia aculeata*, single node; c) *Discocactus horstii*, whole plant; d) *Mammillaria microhelia*, single mammilla; e) *Ceropegia stapeliiformis*, juvenile whole plant; f) *Euphorbia caput-medusae*, distal end of lateral shoot; g) *Gymnocalycium baldianum*, whole plant from above; h) *Euphorbia obesa*, whole plant from above; i) *Lophophora williamsii*, whole plant from above. A: areole. Af: flowering areoles (adult plant body). As: sterile areoles (juvenile plant body). Fs: flower scar. L: leaf. Ls: leaf spine. Lsc: leaf scar. Ph: phylloclade (126). Ps: prophyll spine (66).

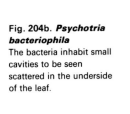

Fig. 204a. *Ardisia crispa*
Bacteria inhabit cavities present in bulges along the edge of the leaf.

Fig. 204b. *Psychotria bacteriophila*
The bacteria inhabit small cavities to be seen scattered in the underside of the leaf.

A domatium (plural domatia), literally a small house, is a cavity within the structure of a plant (stem or leaf or root) (**106, 205b**) which is inhabited by ants, or possibly mites. The morphology of domatia vary considerably. They are formed by the plant even in the absence of the animal (unlike galls **278**) and may be coupled with the production by the plant of some sort of food body or nectary (**78, 80**). The form of the domatium may be simple such as the hollow groove in the adaxial base of the leaf of *Fraxinus* (mites), or the cavity formed by the over-arching of tissue at the junction of two major veins (mites **205c**). Elaborate examples take the form of hollow internodes (**78**) or petioles (**205d**) with entrance holes (ants) or hollow swellings on the under surface of leaves (ants). Ants inhabit the hollowed out woody stipular spines of *Acacia* species (**205a, a¹**). Quite distinct from domatia are cavities in leaves inhabited by bacteria ('leaf nodules') which are found typically in members of the Rubiaceae. Bacteria either accumulate in hydathodes (water excreting glands) on leaf margins (**204a**) or in enlarged substomatic cavities on either leaf surface (**204b**). The bacteria probably invade these cavities during leaf development in the bud where they are associated with mucilage secreted by colleters (**80**).

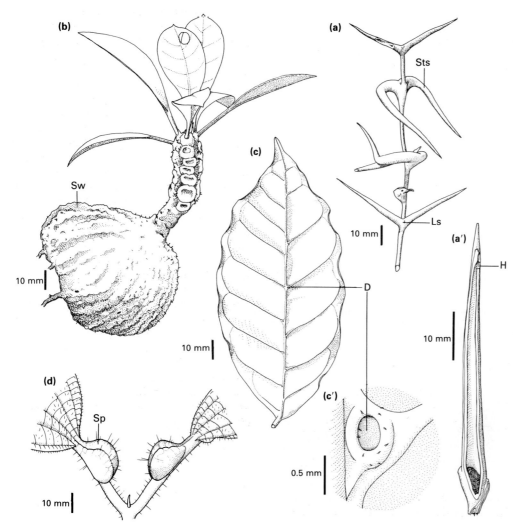

Fig. 205. a) *Acacia sphaerocephala*, series of stipule spine pairs (56); a¹) ditto, section of spine; b) *Myrmecodia echinata*, whole plant; c) *Coffea arabica*, leaf from below; c¹) ditto, single domatium; d) *Tococa guyanensis*, petioles *Sinarundinaria* sp., condensed branching, see **193c**; d) *Crataegus monogyna*, shoot cluster at node; e) *Stachys sylvatica*, flower cluster at node; f) *Forsythia* sp., flower cluster at node; g) *Asparagus plumosus*, condensed shoot system of cladodes. B: bract. Cl: cladode (126). Css: condensed shoot system. F: flower. Fb: flower bud. Fr: fruit. Sl: scale leaf. Ss: shoot spine (124). Vs: vegetative shoot. of pair of leaves. H: entrance hole. Ls: leaf scar. Sp: swollen hollow petiole, entrance holes on abaxial side. Sts: hollow stipular spine. Swo: swollen root containing cavities.

Morphology is the study of shape. The study of plant shape has often been associated with a philosophical attitude. One of the first plant morphologists, Theophrastus (circa 400 BC) was a philosopher and this linkage has persisted throughout time. The history of the subject is detailed by Arber (*The Natural Philosophy of Plant Form*, 1950) and the philosophical attitude by Sattler (1982, 1986). The approach of the German poet and philosopher Goethe (b. 1749) exemplified the recurrent desire of botanists to find a structural identity for plants, something as obvious as the head, tail, and heart of an animal. What is the meaning of a plant, what is its gestalt? Goethe recognized the change in form ('metamorphosis') of leaves in a plant; a study of development will show that the foliage leaf, the sepal, and the petal will each originate from an initially equivalent leaf primordium at the shoot apex. Different structures having the same origin in this manner are said to be homologous (1). Thus the classical interpretation of plant parts, as described by Sachs (1874), recognized four categories of organ each with many homologous variations. These were stem (caulome), leaf (phyllome), root (rhizome—a term now applied only to underground stems, not roots), and hair (trichome). Stem and leaf together constitute a shoot. Plants have been described in terms of alternative structural units more recently (282) but these four basic and usually recognizable morphological categories are universally employed. Nevertheless there are many instances where an attempt to identify the parts of a plant in accordance with this classical framework fail

or become a matter of opinion (e.g. Sattler *et al.*, 1988). The final appearance of a structure does not indicate how it develops and a developmental study will often be helpful in interpreting a particular morphology (e.g. 20, 44). This is especially so when meristematic activity leads to two organs developing as one or remaining connected. This phenomenon may be 'normal' for the plant in question (cacti tubercule 202, epiphylly 74, adnation and connation 234) or result from an abnormal disruption of meristem activity, e.g. fasciation (272)—one form of disruptive development or teratology (270). A danger in the assumption that every morphological feature must be explainable within the classical scheme is that an actual departure

from the 'norm' within the plant kingdom will not be recognized as such or is passed off as an organ '*sui generis*' (literally 'of its own kind', i.e. a one off, 'atypical', an 'inexplicable structure', e.g. 122) if they defy classical interpretation. A more flexible approach is possibly advisable (Groff and Kaplan 1988). Sattler (1974) for example advocates the recognition and expectation of structures that in their development fall between the rigid bounds of leaf and stem (see phylloclades 126). There is no doubt that a few plants are evolving forms that cannot sensibly be accommodated in traditional descriptions (see *Streptocarpus* for example 208). Some of these are outlined here and described as 'misfits' (208–212); misfits, that is, to a botanical

discipline not misfits for a successful existence. Similarly, the haustoria of many parasitic plants do not have a conventional morphology (108) and the leaves of the Lentibulariaceae are difficult to reconcile with the classical mould (Sculthorpe 1967); they are included here under indeterminate growth (90).

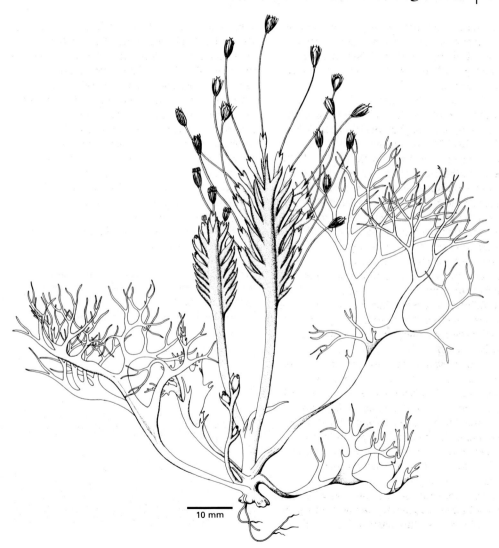

Fig. 207. *Mourera weddelliana* (Podostemaceae) entire plant (210). Redrawn from Tulasne (1852).

10 mm

Most plants in the family Gesneriaceae (dicotyledons) have a conventional morphology, although there is a tendency in many for there to be an inequality in cotyledon size (anisocotyly 32). In some genera (especially *Streptocarpus* subgen. *Streptocarpus* (**209a–f**), but also *Acanthonema*, *Trachystigma*, *Monophyllaea*, *Moultonia*, and *Epithema* **209g, h**) one cotyledon completely outgrows the other and the plant has a growth form that is not compatible with traditional concepts (206). Jong and Burtt (1975) suggest that *Streptocarpus fanniniae* for example would have to be described in conventional terms as follows: 'The plant composed entirely of numerous petiolate leaves (i.e. no stem), the long trailing petioles rooting from the lower surface as they creep over the substrate forming a dense tangled mat. Accessory leaves arising at regular intervals usually on the upper surface of the long petioles and they in turn forming further accessory leaves. Inflorescences developing at the junction of lamina and petiole'. Jong and Burtt (1975) avoid any attempt to describe such a structure by making homological (1)

comparisons with 'ordinary' plants, but identify the basic unit of construction of these plants as a 'phyllomorph', i.e. a leaf blade (or lamina) plus its proximal petiole (or 'petiolode', as it has a much more elaborate morphology than most conventional petioles). When a *Streptocarpus* seed germinates one cotyledon only enlarges and becomes the first phyllomorph of the plant (**209a–d**). The apical meristem of the seedling, which would normally continue development to form the epicotyl, becomes incorporated in the tissue of the upper ('adaxial') surface of the enlarging cotyledon and will be found at the junction of the petiolode and lamina (**208**). This meristem, typical of a phyllomorph, is termed the groove meristem (as it is visible as an elongated depression). In some species the groove meristem gives rise to an inflorescence and the plant, which only consists of the one cotyledonary phyllomorph, dies after fruiting. In other species the groove meristem produces few or many additional phyllomorphs and inflorescences with varying degrees of regularity and second order phyllomorphs may give rise to third order

phyllomorphs and so on. Root primordia form in the under surface of the petiolode. In *Epithema* (Hallé and Delmotte 1973) the single large cotyledon dies in the dry season and it is the third leaf of the plant that forms a fertile phyllomorph (**209g, h**). Phyllomorphs are capable of extended growth over a number of seasons. This is due to the activity of two additional meristems. The basal meristem (**209d**) is situated at the proximal end of the lamina, next to the groove meristem, and continued cell division in this area increases the length of the lamina in favourable conditions. Conversely, the lamina has the ability to jettison its distal end by the formation of an abscission layer across the lamina in unfavourable (dry) conditions. The petiolode can also increase in length due to the activity of a petiolode meristem (**209e**) situated transversely across the petiolode beneath the groove meristem. Elongation at this point not only adds to the length of the petiolode but also breaks the groove meristem into a number of separate regions, termed by Jong and Burtt (1975) detached meristems (**209d**). Each

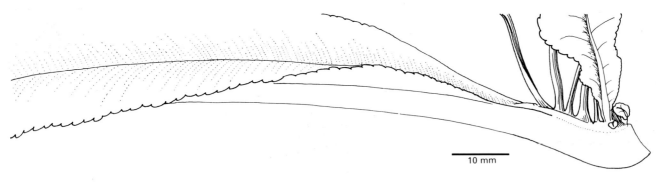

Fig. 208. *Streptocarpus rexii*, base of single phyllomorph with daughter phyllomorphs arising from meristematic region.

10 mm

detached meristem can continue to produce additional phyllomorphs and/or inflorescences. These various phyllomorph characteristics occur with different emphasis in different species. The phyllomorphic construction within the Gesneriaceae is quite unlike the leaf with axillary bud format of most flowering plants and must represent an alternative evolutionary trend (cf. Lemnaceae 212).

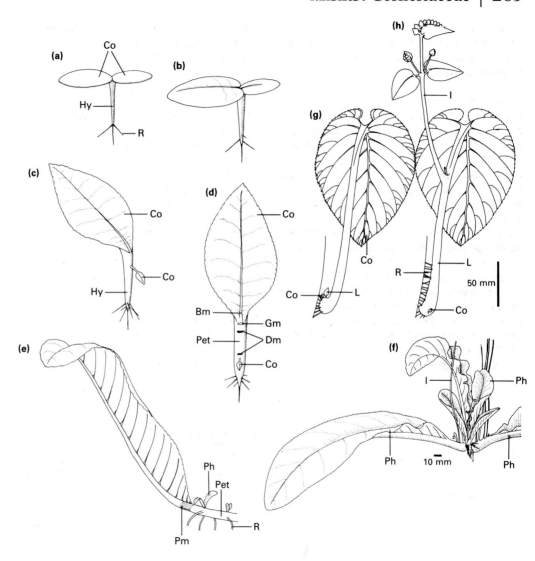

Fig. 209. a–e) *Streptocarpus fanniniae*, stages of development of seedling; f) *Streptocarpus rexii*, base of plant; g) *Epithema tenue*, sterile whole plant; h) *Epithema tenue*, fertile whole plant. Bm: basal meristem. Co: cotyledon. Dm: detached meristem. Gm: groove meristem. Hy: hypocotyl. I: inflorescence. L: leaf no. 3. Pet: petiolode. Ph: phyllomorph. Pm: petiolode meristem. R: root. g, h) redrawn from Hallé and Delmotte (1973), remainder from Jong and Burtt (1975).

Fig. 210.
A 19th century engraving by Thuret (1878) of the thallus of the marine alga *Cutleria multifida*. Compare with Figs. **207** and **211**.

Plants belonging to the two families Podostemaceae (dicotyledons, flowers without perianth segments but enclosed in a spathe 140) and Tristichaceae (dicotyledons, flowers with 3 to 5 perianth segments) live in fast-flowing tropical streams of Asia, Africa, and South America. They vary enormously in morphology (**207, 211**) and in the absence of flowers are not recognizable as Angiosperms. Just as the vegetative body of some parasitic plants (e.g. *Rafflesia* 108) is described as 'mycelial' reflecting its fungal appearance, so the structure of these plants is described as 'thalloid' due to its superficial similarity to that of various algae or liverworts. Some members of these families do give the appearance of having a stem bearing leaves but these structures may merge in their appearance; in some species for example the 'leaves' are indeterminate (90) and retain an active apical meristem. The genera of the Tristichaceae bear non-vascularized scale leaves only. A germinating seed does not produce a radicle, but the hypocotyl bears adventitious roots (98). These roots may then develop into an elaborate structure, fixed to rock surface (a 'hapteron', again terminology borrowed from algae description) and can become dorsiventrally flattened and contain chlorophyll (cf. the roots of some orchids **198**). In common with studies of other plants that have a construction that cannot be reconciled with traditional morphological expectations (206), it is probably pointless to search for homological comparison (1) between these plants and conventional Angiosperms. Nevertheless in theory they are generally supposed to be basically of root origin bearing endogenous shoot/leaf structures in the manner of root buds (178) (Schnell 1967). An extensive bibliography of the Podostemales is to be found in Cusset and Cusset (1988) together with an account of the morphology of members of the Tristichaceae.

Fig. 211. Examples of variation of growth form in the Podostemaceae. a) *Rhyncholacis hydrocichorum*, b) *Marathrum utile*, c) *Castelnavia princeps*. Redrawn from Tulasne (1852). (See also **207**.)

The family Lemnaceae (monocotyledons with close affinities to the Araceae) is composed of four genera: *Spirodela*, *Lemna*, *Wolffiella*, and *Wolffia*. All species are represented by very small aquatic plants floating on or at the surface of fresh water. Each plant consists of either a single 'frond' or 'thallus' or a more or less temporarily connected series of these structures. Fronds vary in size in different species from about 10 mm in length (*Spirodela* species **213e**) to 1.5 mm in length (*Wolffia* species **213b, c**) and are generally flattened distally with a narrow proximal end. Roots are either absent, solitary, or few in number and develop from the under surface. The edges of the frond bear two (usually one in *Wolffia*) meristematic zones each sunken into 'pockets' more or less protected by a flap of tissue. New fronds develop within these pockets from which they sooner or later become detached, forming clones of fronds. In adverse conditions very small 'resting' fronds are produced (turions 172). From time to time a pocket becomes reproductive and male and female flowers consisting of androecium and gynoecium only (**213b**) are formed. *Lemna* clones (e.g. *Lemna perpusilla*) can show considerable symmetry of organization (228): fronds emerge from pockets in strict sequence from one side to the other and clones are either always left- or right-handed depending upon which side produces the first frond. Fronds of left-handed clones will have their reproductive pockets on the right and vice versa. Such small and simple plants do not lend themselves to morphological interpretation even by developmental studies.

Each frond has been considered to consist of a distal leaf lamina plus a proximal narrow region of combined stem and leaf origin together with meristematic zones (and thus similar in concept to the phyllomorph of *Streptocarpus* 208). Studies of the larger species of *Spirodela* and *Lemna* have suggested a conventional arrangement of distichous buds on a very short stem more or less lacking subtending leaves and with either a terminal flattened stem end (cladode 126) or a terminal leaf, the apical meristem being lost.

Fig. 212. *Lemna minor*
Fronds viewed from below.
One root per frond.

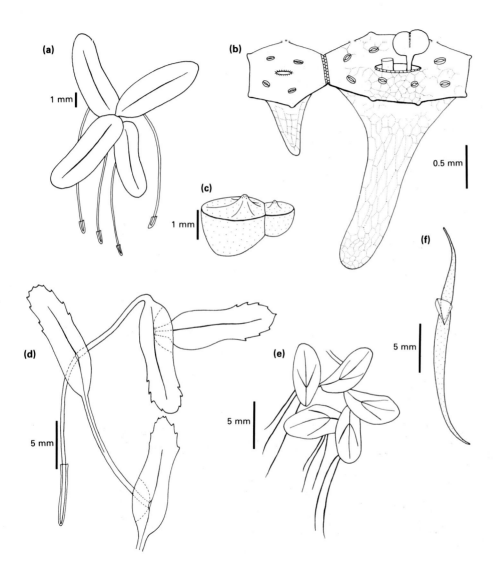

Fig. 213. a) *Lemna valdiviana*, b) *Wolffia microscopia*, c) *Wolffia papulifera*, d) *Lemna trisulca*, e) *Spirodela oligorhiza*, f) *Wolffiella floridana*. Redrawn from Daubs (1965). Whole plant in each case.

Part II

Constructional organization

'What a complex matter in its summation, but what a simple one in its graduated steps, the shaping of a tree is.'

Ward (1909) *Trees: Form.*

'(Actually) we need a solid geometry of tree form to show how systems with apical growth and axillary branching, rooted in the ground and displaying foliage, pervade space.'

Corner (1946) *Suggestions for botanical progress.*

Fig. 215.
The prototype "Raft of the treetops" on its inaugural flight in South America. A well organized construction designed to allow botanists to study, *in situ*, the architectural details of the surface of the tropical rain forest for the very first time.

Fig. 216a. *Pinus* sp. (a Gymnosperm, 14)

Fig. 216b. *Populus* sp.

These botanically accurate computer images were synthesized at the Laboratory of Biomodelization at the Centre de Coopération Internationale en Récherche Agronomique pour le Développement, Montpellier, France. (Reffye *et al.* 1988).

The aspects of plant morphology described in the first part of this guide are to a large extent static features that can be identified in a plant (although sometimes a study of the development of the structure over time is needed in order to understand the final construction). However, a flowering plant is not a static object. It is a dynamic organism constantly growing and becoming more elaborate. Its continued construction is represented by progressive accumulation (and loss) of the morphological features described in Part I. Plants do not grow in a haphazard way but in an organized flexible manner controlled by internal and environmental factors. Part II considers aspects of the dynamic morphology of plants which are not necessarily the features that can be appreciated by studying a plant at one point in its life span. Happily, morphological clues to earlier sequences of events can often be found on a plant such as scars of jettisoned organs or progressive changes in comparable organs of different ages. As a plant grows and becomes more elaborate, it is possible to monitor these sequences at many levels, such as increase in cell numbers, increase in weight, or leaf number and area.

However, greater insight into the developmental morphology or 'architecture' of the plant is revealed by a study of bud activity. New structural components in a plant's framework are developed from buds. A bud develops into a shoot (also termed here a shoot unit 286). The term bud implies a dormant phase but this does not always take place (262). It is therefore more accurate to talk in terms of apical meristems rather than of buds (see page 16 for an introductory discussion of bud and meristem terminology). The contribution of buds or apical meristems to the progressive development of a plant's growth form can be considered under three related headings. Firstly the position of the bud within the plant's framework, secondly the potential of that bud if it grows, i.e. what it will develop into and how fixed is its fate (topophysis 242), and thirdly, the timing and duration of the bud's growth and that of the resulting shoot with respect to the rest of the plant. The morphological consequences of apical meristem position, potential, and time of activity are described in Part II together with an indication of what can go wrong (meristem disruption 270–278) and an example of what can be discovered about the overall morphological entity of a plant (plant branch construction 280–314). An understanding of plant construction necessitates a recognition of the units of that construction (280–286) (plus an awareness that some plants will inevitably not conform to the general flowering plant format 206–212). This section is confined to a consideration of shoot construction (the details to be seen in rooting systems are not yet sufficiently understood 96), and uses advances in the study of tree architecture as an illustration (288–304).

Fig. 217. *Platanus orientalis*. Manipulation of branching pattern by pruning. An example of traumatic reiteration (298).

Phyllotaxis (alternative phyllotaxy) is the term applied to the sequence of origin of leaves on a stem (cf. rhizotaxy 96). The phyllotaxis of any one plant, or at least any one shoot on a plant, is usually constant and often of diagnostic value. In monocotyledons only one leaf is borne at a node although a succession of short internodes interspersed between long may mask this condition. Leaves of dicotyledons are present one to many at a node. The relative positions of leaves on a plant must affect the interception of light and, more importantly, the position of a leaf usually fixes the position of its subtended axillary bud (or apical meristem 16). Thus the phyllotaxis of a plant can play a considerable role in determining the branching pattern of a plant, particularly for woody perennials (288–304). The study of phyllotaxis has led to an extensive terminology and also to a preoccupation with the Fibonacci series (220).

Fig. 218. *Ravenala madagascariensis*
Leaf primordia and therefore leaves develop in two rows on the apical meristem exactly 180° apart giving a distichous phyllotaxis (Fig. **219c**).

Phyllotaxis terminology

A. *One leaf per node*
(sometimes referred to as '*alternate*' in contrast to two per node '*opposite*', see below)
Monostichous. All leaves on one side of stem, i.e. one row as seen from above (**219a**). This is a very rare phyllotaxis and is most often accompanied by asymmetrical internode growth between successive leaves resulting in a slight twist. As a result the leaves are arranged in a shallow helix and the phyllotaxis is termed *spiromonostichous* (**219b, 226**).
Distichous. The leaves are arranged in two rows seen from above, usually with 180° between the rows (**219c, 218**). This is a common condition and is a diagnostic feature of grasses (180). If a slight twist is superimposed on this phyllotaxis, the result is a *spirodistichy* (**219d, 220**).
Tristichous. Leaves in three rows with 120° between rows (**219e**). Typical of the Cyperaceae (196). Twisting may occur—*spirotristichous* phyllotaxis (**219f**).
Spiral. This term is applied if more than three longitudinal rows of leaves are present, e.g. 5 rows as seen from above (**219g**) or 8 rows as seen from above (**219h, 132, 246**). The exact nature of the spiral is described in terms of a fraction indicating the angle between any two successive leaves (220).

B. *Two leaves per node*
Opposite. The two leaves at each node are 180° apart and form two rows as seen from above (**219i**) (the same arrangement often results from internode twisting e.g. **225a**). When successive pairs are orientated at 90° to each other, four rows of leaves will be visible from above and the phyllotaxis is *opposite decussate* (**219j, 233**). In some plants, successive pairs are less than 90° apart and the phyllotaxis is described as *bijugate* (**219k**), leading to a double spiral having two rows of leaves (genetic spirals 220). This arrangement can be referred to as spiral decussate.

C. *Three or more leaves per node*
Whorled. A fixed or variable number of leaves arises at each node. Leaves in successive whorls may or may not be arranged in discrete rows as seen from above. If so, then the whorls are often neatly interspaced (**219l, 229c**). (A *pseudowhorl* results in plants with one leaf per node if series of very short internodes are separated by single long internodes **260a**).

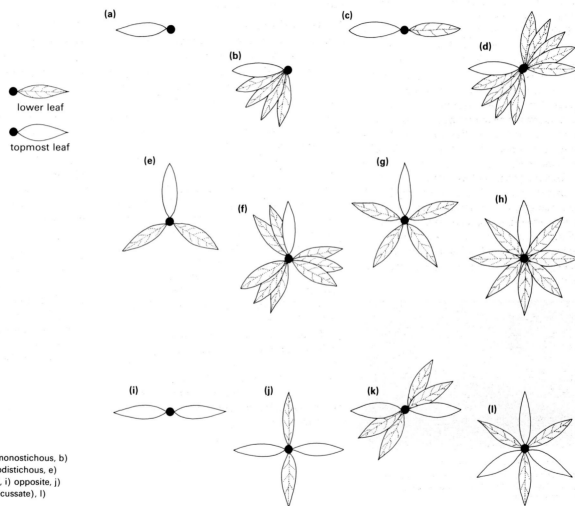

lower leaf

topmost leaf

Fig. 219. Phyllotactic arrangements: a) monostichous, b) spiromonostichous, c)distichous, d) spirodistichous, e) tristichous, f) spirotristichous, g, h) spiral, i) opposite, j) opposite decussate, k) bijugate (spiral decussate), l) whorled.

It is customary to describe the phyllotaxis of plants having one leaf per node (distichous, tristichous, and spiral 218) in terms of a fraction, i.e. $\frac{1}{2}$, $\frac{1}{3}$, $\frac{2}{5}$, etc. This fraction is a measure of the angle around the stem (azimuth) between the

Fig. 220. *Ischnosiphon* sp.
A distichous phyllotaxy (cf. Fig. **219c**) in which a twisting of the internodes takes place after leaf primordium production, resulting in a spirodistichous arrangement (Fig. **219d**). Pulvinus (46) present at the base of each lamina.

points of insertion of any two successive leaves. Thus in $\frac{1}{3}$ phyllotaxis (tristichous **219e**) there is $\frac{1}{3} \times 360 = 120°$ between two longitudinally adjacent leaves, in $\frac{2}{5}$ phyllotaxis there is $\frac{2}{5} \times 360 = 144°$ between two successive leaves (Fig. **221a**). An imaginary line can be drawn spiralling around such a stem which passes through the point of attachment of each next youngest leaf in turn. This is termed the *genetic spiral* (**221a, b**, Introduction).

An estimate of the phyllotactic fraction can be found by following the genetic spiral around the stem from any one older, lower, leaf to the 1st younger leaf directly in line above it. Leaves seen to be arranged in a common longitudinal line are said to lie on the same orthostichy. A distichous plant (**219c**) will have two orthostichies, a tristichous plant three (**219e**), and the example of $\frac{2}{5}$ spiral phyllotaxis (**221a**), five orthostichies. In Fig. **221a** the lower leaf will be given the number 0 and the leaf arrived at vertically above it will be found to be number 5. The genetic spiral will have been found to have passed twice around the stem giving a fraction of $\frac{2}{5}$ and hence an indirect measure of 144° between any two successive leaves.

In Fig. **221b** the phyllotactic angle is 135° ($\frac{3}{8}$, i.e. leaf 8 is above leaf 0 and reached by passing three times around the stem). The ease with which this measurement can be made may be more or less confounded by the amount of internode twisting or leaf primordium displacement that has taken place as developing leaves become shifted away from initially precise orthostichies (230).

The phyllotactic fractions almost invariably found in plants with spiral phyllotaxis are:

$$\frac{1}{2} \ \frac{1}{3} \ \frac{2}{5} \ \frac{3}{8} \ \frac{5}{13} \ \frac{8}{21} \ \frac{13}{34} \ \cdots \cdots$$

which represent angles of:

$$180° \ 120° \ 144° \ 135° \ 138°28' \ 137°6' \ 137°39'.$$

It will be seen that both the numerators and denominators form Fibonacci series, i.e. each successive number is the sum of the preceding two ($2 + 3 = 5$, $3 + 5 = 8$, etc.). This series of fractions based on Fibonacci numbers could be continued indefinitely and each successive fraction would represent an angle approaching nearer and nearer to, but never actually reaching, 137° 30′ 28″. This angle represents the sector of a circle with significant properties.

In Fig. **221c** the ratio of A to B is the same as the ratio of B to the whole of the circumference (A + B). Thus, for example, if A = 1 then B will be 1.61803 . . . , or if B = 1 than A + B will = 1.61803 . . . . 1.61803 . . . is a number with no finite value, an irrational number (it is represented by

$$\frac{1 + \sqrt{5}}{2}$$

and is termed Phi, φ; a more familiar irrational number is Pi, $\pi \frac{22}{7}$). Any line or circle divided by this ratio of 1 to Phi is said to be divided by the golden ratio and these proportions are invariably found to be pleasing to the eye.

(Continued on page 222.)

(a)

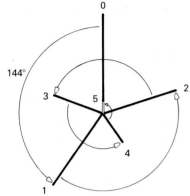

144°

(b)

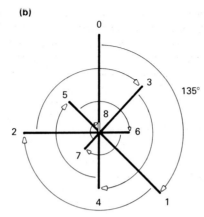

135°

Fig. 221. a) $\frac{2}{5}$ phyllotaxis, the arrows follow the genetic spiral. Positions 0 and 5 lie on the same orthostichy. b) $\frac{3}{8}$ phyllotaxis, c) a circle divided by the golden ratio A=1, B=1.61803 . . . , d) the relationship of the golden ratio with the creation of a logarithmic spiral.

(c)

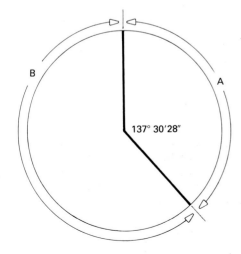

B

A

137° 30′28″

(d)

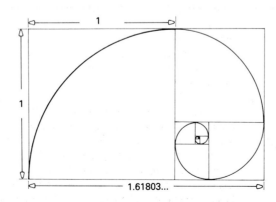

1

1

1.61803...

Fig. 222. ***Opuntia*** sp.
Surface view of a phylloclade (126) showing rows of areoles (202) in logarithmic spirals. Flower buds are developing at some sites.

The presence of the Fibonacci series and hence the golden ratio in the phyllotaxis of plants (220) has led to much investigation and many explanations. The two most relevant points are as follows:

1. If the leaves (and hence subsequently branches) of a plant were spaced up the stem at intervals of exactly $137° \ 30' \ 28''$ then *no* leaf or branch would be positioned exactly above another, which has implications for the shading of one leaf or branch by a higher one. Spirals with phyllotactic fractions of $\frac{5}{13}, \frac{8}{21}, \frac{13}{34}$ and above are approaching this angle.

2. The golden ratio has a relationship with a logarithmic spiral. A simple visual illustration of this is shown in Fig. **221d**.

A logarithmic spiral (or helix) can be extended indefinitely outwards or inwards and is therefore always of the same shape regardless of its dimensions. The shell of a snail, and of *Nautilus*, forms such a spiral. As the animal increases in size it occupies a progressively larger volume. However, both the animal and its shell retain the same *shape* regardless of size. A similar growth phenomenon takes place at the apical meristem of a plant when leaf primordia of initially small size develop but of necessity occupy the same proportion of the apex surface (**18**). The consequence of this packing of enlarging organs can be seen on a pineapple fruit (**223**) or on the inflorescence head of a sunflower (*Helianthus* spp.). All the sunflower seeds are the same shape but not the same size. Furthermore, they are arranged in radiating spiral rows; two directions of rows are visible, one set clockwise and one set anticlockwise. These rows are termed parastichies and form logarithmic spirals. All the spaces between these intersecting logarithmic spirals are the same shape regardless of size.

Developing leaf primordia enlarging at a growing shoot apex similarly continue to fit comfortably together as they expand in basal area and will inevitably form two sets of interlocking parastichies in the process (**223**). This uniformity of shape resulting from logarithmic spirals does not occur unless the number of parastichies in each direction conforms to the Fibonacci series. Thus counts of rows on sunflower head, or pineapple fruit, conform to the following series:

   1   2   3   5   8   13   21   34 etc. in one direction

   2   3   5   8   13   21   34   55 etc. in the other direction.

Intermediate combinations do not occur and would result in distorted structures. This series is complementary to the series giving a measure of the angle between any two successive leaves on the genetic spiral as it gives a measure of the angle for sector B of the circumference rather than sector A (**221c**). An extended account of Fibonacci and phyllotaxis is given in Stevens (1974).

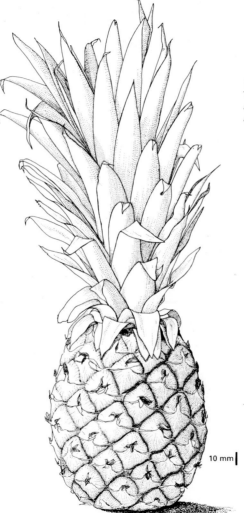

Fig. 223. *Ananas comosus*, fleshy multiple fruit (**157m**). This specimen has eight parastichies running upwards and anticlockwise, and thirteen parastichies running upwards and clockwise.

10 mm

The standard terminology describing leaf arrangement (phyllotaxis 218) is usually straightforward in its application. Nevertheless, in many cases the phyllotaxis is ambiguous or confusing. Examples are given here under the general heading of phyllotactic problems although the problem is for the morphologist, not for the plant. Departures from the customary arrangement of leaves fall into two main categories: (a) plants in which an initially common phyllotaxis is masked by secondary shifts of orientation (**219b, d, f**) and (b) plants showing departure from the common types. More than one type of phyllotaxis can occur on the same plant. This is often the case in woody plants having both orthotropic and plagiotropic shoots (**246**). A change in phyllotaxis can occur along a single shoot again often reflecting a change of orientation (e.g. metamorphosis 300). A change from opposite decussate to spiral arrangement is common in dicotyledons where the cotyledons are invariably an opposite pair. The portion of axis between two discrete phyllotaxes will itself have a transitionary and confused leaf positioning (**227**). The phyllotaxis along a shoot can also be modified if growth is rhythmic (**260**) and a series of small cataphylls of a resting apical bud with very short internodes is interspersed between a series of leaves with large bases and with long internodes (**119g**). A strict 90° opposite decussate arrangement along one section of stem may shift a few degrees at the point at which a dormant bud was sited. A clockwise spiral phyllotaxis (as seen from above) following the genetic spiral from an older to a younger leaf can switch to a counter-clockwise phyllotaxis and vice versa. The initial phyllotaxis may be lost or confused by a displacement due to subsequent meristematic activity. A distichous arrangement of leaves can become spirodistichous as internodes expand and lengthen (**220**). Successive internodes can twist through 90° converting an opposite decussate phyllotaxis into an apparently opposite one (**224**). This takes place particularly on plagiotropic branches (**225a**). Similarly an opposite decussate origin can be converted into a bijugate, spiral decussate arrangement (metamorphosis 300).

(Continued on page 226.)

Fig. 224. *Eugenia* sp. A plagiotropic (246) branch having opposite decussate phyllotaxis (Fig. **219j**) in which all the leaves are subsequently brought into the horizontal plane by twisting of internodes, plus adjustment by pulvini (46).

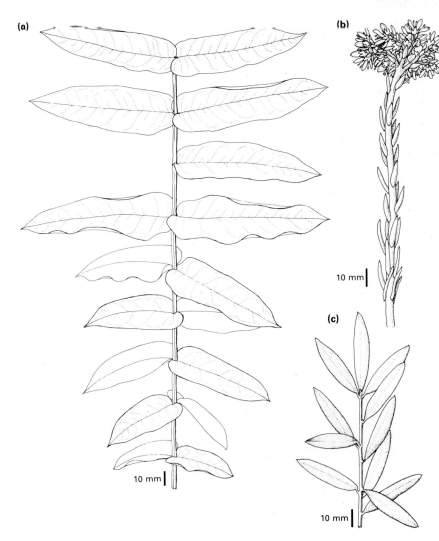

(a)

(b)

10 mm

(c)

10 mm

10 mm

Fig. 225. a) *Eucalyptus globulus*, horizontal shoot from above. Opposite decussate (**219j**) phyllotaxis at proximal end, opposite (**219i**) phyllotaxis at distal end. b) *Sedum reflexum*, haphazard spiral phyllotaxis with no discernible geometry. c) *Olea europaea*, variable internode lengths giving an undeterminable phyllotaxis.

Fig. 226. *Costus spiralis*
A spiromonostichous phyllotaxis (cf.
Fig. **219b**).

A perfectly normal spiral phyllotaxis at the shoot apex can be lost completely by stem/leaf adnation (234) as the system develops. This is typical of many members of the Solanaceae (**234a, b**). The position of a leaf in the phyllotactic sequence is very occasionally occupied by a shoot (usually represented by a flower as in some members of the Nymphaeaceae). If a very small scale leaf (bract 62) subtends the flower the 'problem' is simply a matter of the relatively large size of the pedicel (or its scar) compared to that of the insignificant bract. If the bract is absent then the flower appears to occupy a leaf site. A precocious vegetative bud developing in this manner could give the appearance of a true dichotomy (258). Unusual phyllotaxes are initiated as the norm at the shoot apices of some species. *Costus* species have a phyllotaxis described as spiromonostichous. Leaf primordia are produced one at a time at the developing apical meristem with an unusually long delay between the appearance of successive leaves (i.e. a long plastochron 18). Each primordium is situated only a few degrees around the stem apex from the last leaf primordium and leaves thus lie on a very gentle helix that does not fit anywhere in the series of spiral phyllotaxes normally found (**226**). The aerial shoots of *Costus* represent the distal end of the sympodial rhizome (**131b**) system and the direction of each helix (clockwise or anticlockwise) changes with each successive sympodial unit. Other unique phyllotaxes are also occasionally encountered. In *Nelumbo* leaves are present on the rhizome in sets of threes. A ventral scale leaf is followed by a dorsal scale leaf and then by a dorsal foliage leaf. In *Anisophyllea disticha* two leaf sizes (dimorphism 30) follow each other in precise order, two to one side, two to the other side as follows: left small, left large, right small, right large, left small, etc. Large leaves are borne on the underside of the horizontal shoot, small leaves on the upper side (Vincent and Tomlinson 1983). A similar phyllotaxis has been noted for *Orixa japonica*, *Lagerstroemia indica*, and *Berchemiella berchemiaefolia* and termed 'orixate'.

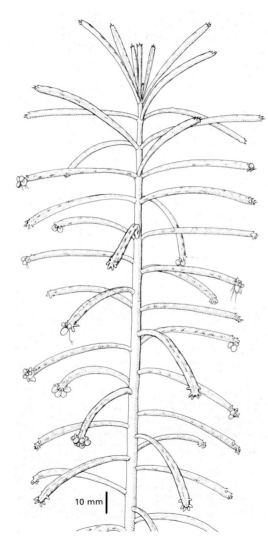

Fig. 227. *Bryophyllum tubiflorum*. Terete (86) leaves bearing adventitious buds (74, 232) distally. The leaves are arranged spirally at the base of the stem and are whorled at the upper half (cf. **233**).

10 mm

Symmetry is an obvious feature of many leaves (26) and flowers (148), but it can also be applied as a concept to whole plants or parts of plants (cf. inflorescences 142). Geometric symmetry is self evident in the case of many cacti and other succulents with limited branching (202). Generally speaking the less the branching the more symmetrical a plant is likely to appear. Symmetry results from a repetition of similar branching constructions (paracladia 142). If the plant produces a series of buds (apical meristems 16) with fixed potentials (242) and these buds are precisely located in association with the subtending leaves which are themselves precisely located (phyllotaxis 218), then a branched structure will develop with an obvious symmetry of pattern. Subsequent damage, unequal bud activation, or growth due to environmental gradients (especially directions of light intensity), innate branch death (244), and activation of buds out of synchrony (e.g. reiteration 298) will distort symmetry in the older plant. Symmetry is particularly apparent in plants with opposite decussate phyllotaxis. The branches developing from bud pairs frequently grow to produce mirror image branch systems (228). On a vertical axis such symmetry may be apparent in three dimensions. In plants exhibiting anisophylly branching at individual nodes will be asymmetrical but these asymmetries themselves will be located in a symmetrical fashion within the plant framework (33e). Such features are usually most readily identifiable and recorded by means of condensed and simplified diagrams especially 'floral' type diagrams (9, 259).

Paradoxically symmetry in a plant often results from the organized location of asymmetrical features. Figure. 229c shows the arrangement of whorled leaves and branches at a node on a vertical axis of *Nerium oleander*. Three branches radiate out obliquely and symmetrically at each node. Each bears a pair of prophylls (66) at its first node followed by a whorl of three leaves at the next node. The three sets interdigitate neatly. Figure. 33d shows the occurrence of two simple leaves (which soon fall) within the complex of pinnate leaves typical of the tree, *Phellodendron*. Compound leaves at these two locations would be forced to occupy the same congested space. A type of symmetry can occur in the form of repeated sequences of organs along an axis. The locations of stem tendrils in virginia creeper

(*Parthenocissus*) provides an example (229b, 310). Tendrils are apparently leaf opposed (122), occur in an infallible left/right sequence, and every third node is tendrilless. Symmetry can apply to the direction of spiral phyllotaxis in successive sympodial units (see *Costus* 226) or the regular sequences of structures along sympodial axes (*Carex arenaria* 235a–c). The sequence of parts on an axis may be asymmetrical rather than repetitively symmetrical but nevertheless repeated predictably between axes (*Echinodorus* 132). Symmetry within a plant can be a fundamental aspect of its architecture apparent even in large trees (304). It is often most noticeable in compact forms of inflorescences (8).

Fig. 228. *Rothmannia longiflora*
A bifurcating branch system (false dichotomy 258) in which the left and right sides are themselves asymmetrical, but form mirror images, the whole system being symmetrical.

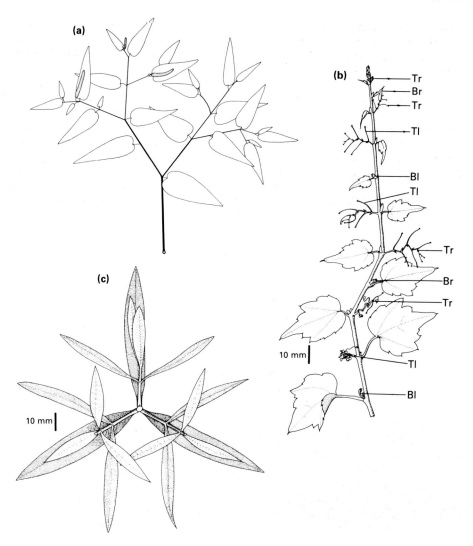

(a)

(b)

Tr
Br
Tr

Tl

Bl

Tl

Tr

Br

Tr

Tl

Bl

10 mm

(c)

10 mm

Fig. 229. a) *Piper nigrum*, an apparently irregular small-scale occurrence of pseudodichotomy (**259d**) results in a regular branching pattern at a larger scale. b) *Parthenocissus tricuspidata*. Tendrils are apparently leaf opposed (122). Every third node has a bud but is tendril-less. Any specimen will fit somewhere into the following sequence: bud left (Bl), tendril left (Tl), tendril right (Tr), Br, Tr, Tl, Bl, Tl, Tr, Br, Tr, Tl, etc. c) *Nerium oleander*, viewed from above, leaves in whorls of three except for the offset pairs of prophylls (66).

Fig. 230a. *Fuchsia* cv. Mrs. Popple

A teratology (270) in which a bud has
become displaced during development and
now occupies a position away from its
subtending leaf axil. (The same specimen has
some abnormal flowers, see Fig. 270.) Such
displacement occurs normally in many plants
(**231b**).

Fig. 230b. *Datura cornigera*

Shoot displaced sideways on to leaf base.
This genus has a very flexible morphology in
this respect (see **234a, b**).

Each leaf of a flowering plant can usually be
expected to subtend a bud (4). The bud is distal
to i.e. above the leaf and subsequently the shoot
into which the bud might develop is distal to the
scar left after the leaf has fallen. The mid-line of
the bud is typically on the same radius as the
mid-vein of the leaf. In many instances, however,
these typical conditions do not hold. The bud
may be absent, the leaf may be absent
(commonly in inflorescences). The bud may
apparently take the place of the leaf (226), more
than one bud may be present in the axil of the
leaf (accessory buds 236) and then all but one
bud may be displaced away from the 'normal'
position (**236a, b**). Where there is one bud it may
be displaced around the axis and then be referred
to as exaxillary. Such displacement is not so
obvious in monocotyledons as the leaf itself
encircles most or all of the stem circumference
(**183d, e**). As the apical meristem develops and
elongates, bud and subtending leaf can become
separated by intervening tissue (adnation **231b**).
In extreme cases the bud will appear to be
associated with the next youngest leaf but to be
on the 'wrong' side of the stem (**231a, e**). Very
occasionally, as in *Thalassia* (Tomlinson and
Bailey 1972), the bud is truly leaf opposed and is
borne at the apex on the 'wrong' side of the
stem. More often a bud or shoot appears at first
to be leaf opposed but is actually at the
termination of a sympodial unit, the main axis
being continued by an axillary shoot (**251a**). A
bud can also be displaced out on to its
subtending leaf rather than up the main stem
axis (**230b**). The leaf will then appear to have no

associated axillary bud (unless there is an accessory bud 236), but will have a bud (typically a flower or inflorescence) somewhere on its petiole or lamina (epiphylly 74). Buds occasionally develop in locations in the complete absence of subtending leaves (adventitious buds 232).

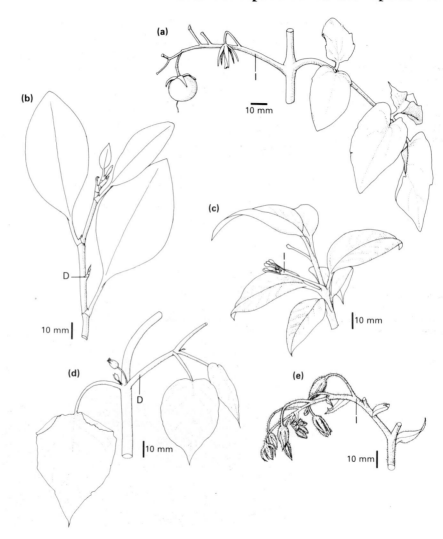

Fig. 231. a) *Lycopersicon esculentum*, the inflorescence, left, developed initially in the axil of the next leaf below; b) *Griselinia littoralis*, bud displaced distally away from its subtending leaf; c) *Hoya multiflora*, inflorescence displaced around the stem circumference; d) *Physalis peruviana*, vegetative shoot displaced upwards away from its subtending leaf and now located opposite an upper leaf which has its own normal bud; e) *Borago officinalis*, as a). D: displaced shoot. I: inflorescence.

A bud is said to be adventitious when it is found in an unusual place (cf. adventitious roots, i.e. especially roots on stems 98). It must be stressed that except in the case of a 'mistake' (teratology 270) by the plant, the so-called unusual location of the adventitious bud is unexpected for the observer, but normal for the plant. Customarily a bud is located in the axil of a leaf, i.e. just distal to the point of attachment of the leaf to the stem. Buds developing in this position (there may be more than one 236) can become displaced away from their subtending leaf by subsequent meristematic activity (230). The term adventitious is applied to a shoot meristem developing anywhere on the plant in the total absence of a subtending leaf (**232**) (it excludes, however, the components of inflorescences in which bracts, i.e. leaves, are often absent). Thus shoots formed on roots (178) arise from adventitious buds. A number of tropical trees are called sapwood trees, having no dead heartwood in the trunk. If the trunk of one of these trees is severed, shoot meristems can develop by resumed meristematic activity of living cells in the centre of the trunk (Ng 1986). These form adventitious buds. Similar endogenous activity can give rise to epicormic and cauliferous shoots (240). Adventitious buds are found on the hypocotyl of a number of plants (**167e**). Another category of adventitious bud is that formed on a leaf (petiole and/or lamina). This situation is referred to as epiphylly (74), and in some cases can be accounted for by axillary bud displacement; in others it represents meristematic activity of groups of cells typically at leaf margins (**233**).

Such adventitious buds are often easily detached, root by means of adventitious roots (98), and may take the form of bulbils (172). Similar adventitious buds form from the broken base of detached leaves of many succulent plants. A bud may appear at first sight to be adventitious if it persists in a dormant state long after all traces of the subtending leaf have disappeared. This is true in some instances for epicormic buds (240) present at the surface of a trunk or branch.

Fig. 232. *Medeola virginiana* Excavated sympodial rhizome. An adventitious bud which is not subtended by a leaf is always present at the proximal end of each sympodial unit (at the extreme right in this view). The direction of growth of this bud, left or right, alternates on successive units and thus forms a predictable branching pattern that may however be more or less disrupted depending upon the environment (Bell 1974; Cook 1988).

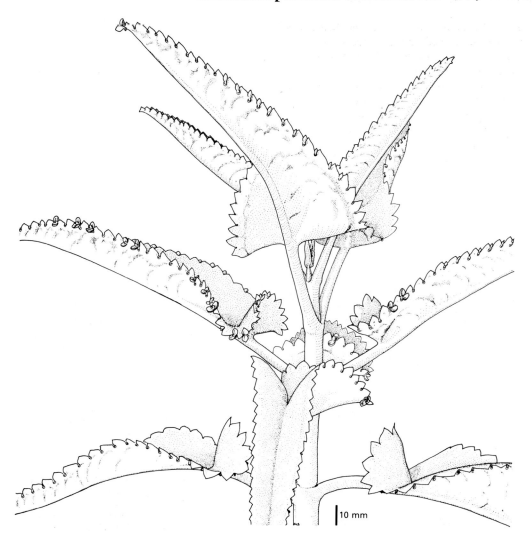

Fig. 233. *Bryophyllum daigremontiana*, detachable buds with adventitious roots are located along the leaf edges (74). Phyllotaxis opposite decussate (cf. **227**).

10 mm

Fig. 234a, b. *Datura cornigera*
Two stages of the development of a side shoot. a) The leaf base and axillary shoot are beginning to grow out on common tissue. b) This process has continued and a long portion of stem and leaf tissue now has the leaf lamina at its apex together with a terminal flower.

The term concrescence is applied to the situation where two structures are united together. If the two (or more) organs are of the same type (e.g. both petals) then they are said to be connate; if the two organs are of different categories (e.g. stamen and petal) they are said to be adnate, although this second term is frequently used for either condition. Both adnation and connation frequently occur in the flower (involving bracts, or sepals, or petals, or stamens, or carpels, 146). Opposite or whorled leaves may be connate at a node (**235f**). Adnation other than in the flower usually involves the fusion of a bud (or the proximal portion of the shoot into which it has developed) with either the subtending leaf (epiphylly 74) or with the adjacent stem (bud displacement **231b**). Basically the concrescent condition arises in two ways. The initially separate organs can actually become firmly attached as they develop side by side at the primordium stage (postgenital concrescence). This happens in the case of carpel connation; it can also occur as a teratology (270). Alternatively, the two organs (either similar or different) are only separate at the very earliest stages of development, becoming and remaining united by the meristematic activity of common tissue on which they are both located. An adnate bud becomes 'carried out' on to its subtending leaf petiole (**230b**), or 'carried up' the stem away from the leaf axil (**230a**). The two parts appear to be fused together but in reality they have developed as one throughout; this is referred to as congenital (or 'phylogenetic') concrescence (adnate or connate). A particularly elaborate example occurs in many rattans (climbing palms 92) where the stalk of the inflorescence (or flagellum) developing in the axil of a leaf, is adnate to the main stem axis right up to the next node above and in some cases out on to the side or underside of the leaf at that node. Some species of *Carex* (e.g. *C. arenaria* **235a–c**), are described as monopodial (250). Only close study of the location of leaves and internal vascular anatomy reveals that the rhizome is in fact sympodial with connation of daughter/parent internodes at exactly repeated intervals.

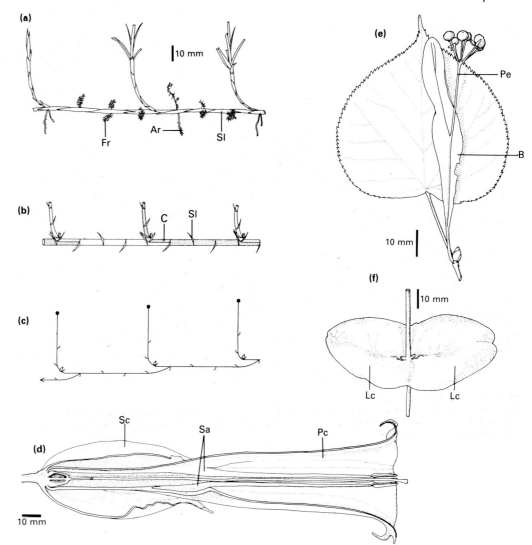

Fig. 235. a) *Carex arenaria*, rhizome with aerial shoots; b) diagram of 'a', stippled portion equals one sympodial unit, the proximal internode of which is connate to one internode of the previous unit; c) stick diagram of 'b' emphasizing consistent organization of components; d) *Datura sanguinea*, half flower (150), sepals are connate, petals are connate, stamens adnate to petals; e) *Tilia cordata*, inflorescence peduncle adnate to subtending bract; f) *Lonicera × brownii*, pair of leaves connate at node. Ar: anchoring root. B: bract. C: connation. Fr: fibrous root. Lc: connate leaf. Pc: connate petal. Pe: peduncle. Sa: adnate stamen. Sc: connate sepal. Sl: scale leaf. a–c) redrawn in part from Noble *et al.* (1979).

It is not uncommon to find more than one bud in the axil of a single leaf (including a cotyledon 162). Usually one bud is more prominent than the other accessory (or supernumerary) buds and will be the first or only bud normally to develop.

A very similar arrangement can result by the formation of a condensed branching system with very short internodes developing from a solitary axillary bud (proliferation 238). Accessory buds may occur side by side in the leaf axil (collateral buds) or may form a row in line with the stem axis (serial or superposed buds). In any one species there is usually a hierarchy in terms of bud size and the sequence in which the buds are able to become activated (**237e–j**). In some instances all the buds in the leaf axil could eventually grow to form the same type of shoot. For example in the basal part of *Eucalyptus* spp. that form lignotubers (**138a**), the main bud of each set of three develops into a shoot. If this is damaged by frost or fire the two accessory buds, one above and one below, will grow to form replacement shoots. Indeed the meristematic region in the axil of a *Eucalyptus* leaf is in some cases capable of the continual production of accessory buds (**237d**). Alternatively, each accessory bud has its own distinctive potential. In various tree species, *Shorea* for example, one of a pair of buds will always grow as soon as it is formed and form a horizontal shoot (syllepsis 262 and plagiotropy 246) whilst the second bud will have a delayed action and can only produce a vertical shoot (prolepsis 262, and orthotropy 246). Various combinations of structures can thus be found in the axil of a single leaf, each organ derived from one of a set of accessory buds, for example there may be a spine and vegetative shoot (**236b**), or an inflorescence and a vegetative shoot, or a flower, a shoot tendril, and a vegetative shoot, as in the case of *Passiflora* (**237c**) (cf. 122).

Fig. 236a. *Leucaena* sp. Accessory buds in the axil of one leaf, all developing as inflorescences.

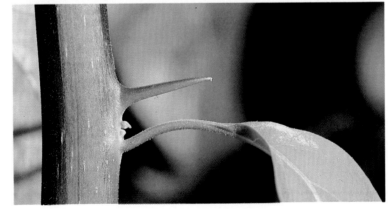

Fig. 236b. *Bougainvillea glabra*
A serial sequence of accessory buds in the axil of one leaf. The upper bud can develop into a stem spine (124), as here, or into an inflorescence (**145d**).

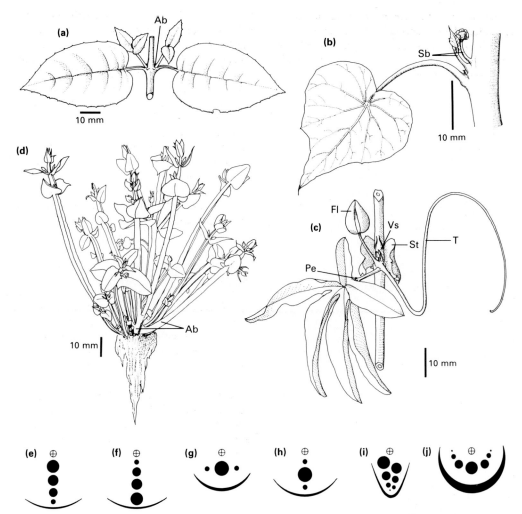

Fig. 237. a) *Fuchsia*, sp., single node; b) *Stephania* sp., single node; c) *Passiflora caerula*, single node; d) *Eucalyptus globulus*, epicormic branching (240) on trunk; e, f, h) serial buds, g, j) collateral buds, i) staggered serial buds. Ab: accessory bud. Fl: flower. Pe: petiole. Sb: serial buds. St: stipule. T: tendril. Vs: vegetative shoot.

When several structures (vegetative shoots, flowers, tendrils, or spines) develop from the axil of a single leaf, they either represent the activity of a set of accessory buds (236), or a condensed branching system with very little internode elongation. In the latter case (proliferation) the single axillary bud has borne its own buds and in turn they have borne their own buds in continual sequence. In this situation each bud or shoot will be in the axil of a different leaf whereas accessory buds are all subtended by the same leaf (**237b**). In some instances very careful study of the younger stages of development down to the primordium stage may be needed to distinguish these two morphologies, possibly supported by the identification of vascular linkages (i.e. does one bud link into another bud or directly into the main stem?) For example, the pair of buds in the axil of a leaf of *Gossypium* (cotton) are frequently described as accessory buds (236), but Mauney and Ball (1959) show them to be proliferation buds by anatomical investigation. The buds within a proliferation shoot complex may all be of the same potential type (flowers **239a, e**; cladodes **239g**; spines **124b**) or may have different potentials and times of activity just as is the case for accessory buds (**239d**). The areole (202) of the Cactaceae is a particularly familiar example of this condition.

Fig. 238. *Ophiocaulon cissampeloids*
The vertical shoot to the left is hanging downwards so that the leaf is above its axillary shoot. The lamina is reoriented by a twisting petiole. This shoot is a stem tendril (cf. Fig. **145b**) and apparently bears a daughter shoot, an inflorescence, at its extreme proximal end, but see explanation in section 122.

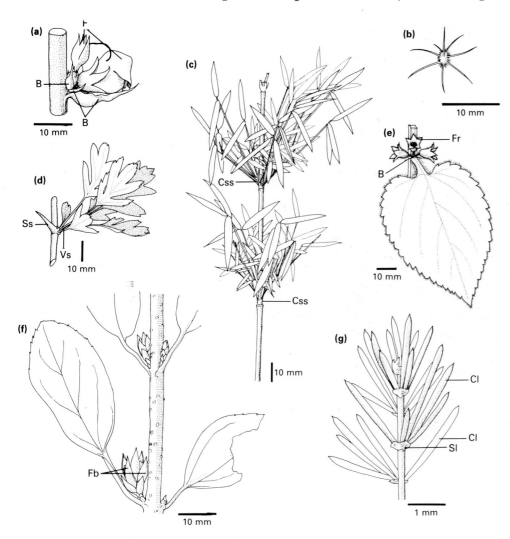

Fig. 239. a) *Verbascum thapsus*, flower cluster at node; b) *Melocactus matanzanus*, single areole (202); c) *Sinarundinaria* sp., condensed branching, see **193c**; d) *Crataegus monogyna*, shoot cluster at node; e) *Stachys sylvatica*, flower cluster at node; f) *Forsythia* sp., flower cluster at node; g) *Asparagus plumosus*, condensed shoot system of cladodes. B: bract. Cl: cladode (126). Css: condensed shoot system. F: flower. Fb: flower bud. Fr: fruit. Sl: scale leaf. Ss: shoot spine (124). Vs: vegetative shoot.

Fig. 240. *Parmentiera cerifera*
Cauliflory. One flower and two fruits.
Each is developing from a persistent
bud complex.

Cauliflory refers to the phenomenon of flowers (**240**) (and subsequently fruits **240, 241**) issuing from the trunk or branches of a tree. The latter condition is also termed ramiflory. The production of single or more commonly, whole clusters of vegetative twigs at discrete locations on the trunk (or occasionally branch) is referred to as epicormic branching (**237d**). The sites of epicormic branching and of cauliflory are often recognizable as swellings or disruptions in the general bark surface. The buds giving rise to cauliflorous or epicormic branching have two distinct origins, either one or both being found in any given species. The buds may be truly adventitious (232) arising endogenously (i.e. deep in existing tissue as does a root primordium 94) by resumed meristematic activity of living cells, the shoots so formed growing outwards to become located in the bark of a tree. Stump sprouts can form similarly on a severed trunk, especially from the cambium zone (but also from sapwood 232). Alternatively, a bud formed on the young trunk or a branch of the tree in the normal manner in the axil of a leaf will remain alive growing outwards a very short distance each year. It thus keeps pace with the gradually expanding trunk, rather than becoming encroached upon by the tree's secondary growth as does a dead branch stump (forming a knot in the wood) or an embedded nail or staple. Such a bud will usually only produce scale leaves but the buds in their axils may also develop and extend outwards alongside the first bud. This process can be repeated until a mass of buds is present at the surface of the bark. Potential epicormic or cauliferous buds formed in this exogenous manner have been termed preventitious in contrast to adventitious buds which have an endogenous origin. The latter also track outwards and multiply in number. Neither type can be described as dormant, as they grow a short distance each year in the manner of a short shoot (254). They are referred to as suppressed buds until such time as an epicormic branch or flower/inflorescence is produced.

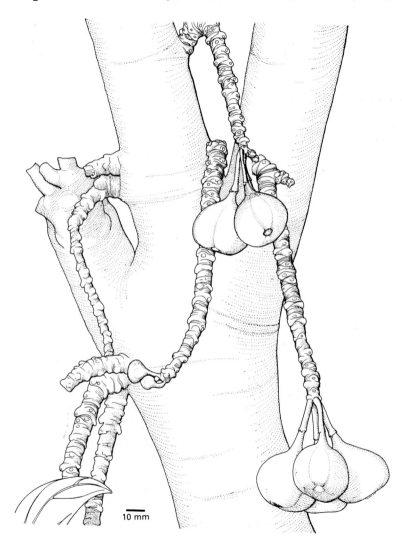

Fig. 241. *Ficus auriculata*. Fruits (syconium **157n**) developing from short shoots on trunk.

10 mm

The framework of a plant is built up from a number of shoots each derived from a bud or apical meristem (16). Some shoots will be of a temporary nature, being shed sooner or later (cladoptosis 268), some buds will remain dormant, some buds and shoots will have a totally predictable and fixed fate with distinctive morphological features, others may have a flexible potential depending upon the experience of the individual bud. In many cases, the inflexible potential or status of a bud or shoot can be demonstrated by detaching and rooting the shoot whereupon it will retain its specific morphology. This irreversible retention of characteristics is termed topophysis and is usually applied to two particular types of phenomenon. One applies to the situation where a plant, typically a tree, has two (or more) types of branch each with different details of leaf shape, flowering ability, and orientation (e.g. plagiotropic or orthotropic 246). Each will retain its individual characteristics on being rooted as a cutting: a fan branch (plagiotropic) of *Theobroma* (cocoa), if rooted, will grow horizontally along the ground and cannot form an upright stem (98). The second form of topophysis applies to shoots of different age categories within the plant. A plant may be described as having juvenile and adult foliage types for example. As the plant grows its passes from a 'juvenile' form to an 'adult' form (**203c**) (age states 314). The actual calendar age of a plant is largely irrelevant in this context. A 'seedling' forest tree may mark time for decades putting on little growth each year until it has an environmental opportunity,

such as a light gap, for developing into an adult tree with different morphological characteristics. Again, material taken from the juvenile or adult parts of the plant may retain their characteristics without change. A frequently quoted example is that of juvenile ivy, *Hedera helix* (monopodial, distichous phyllotaxy, rooting, climbing by adventitious roots, flowerless) and adult ivy (sympodial, spiral phyllotaxy, lacking roots, flowering). The artificially propagated juvenile form will progress to the adult phase, but artificially propagated adult ivy will grow as a shrub and cannot normally revert to the monopodial climbing phase. In a sense it could be argued that topophysis applies to all buds (apical meristems) of all flowering plants. The potential of the bud is either inflexible, for example it can only form an inflorescence and it will always grow into this particular structure, or the

potential of the bud is variable but again the outcome of its growth is predictably dependent upon its precise location and time of appearance within the elaborating branch construction of the plant (see architectural models 288, reiteration 298, metamorphosis 300). Thus, of the several buds in the axil of a *Bougainvillea* leaf one has the potential to develop into a vegetative shoot if conditions are suitable, another will always form either a determinate spine or a determinate inflorescence (**145d, 236b**). Artificial rooting of such structures would doubtless confirm the topophytic nature of these potentials. The term topophysis should properly be retained for situations in which the type of growth of a meristem is irreversibly fixed although this represents only the more obvious example of a general phenomenon of controlled meristem potential.

Fig. 242. *Gleditsia triacanthos*
A young epicormic shoot emerging from the trunk of an old tree. This location causes the formation of a determinate stem spine rather than a new vegetative branch or a flower (cf. cauliflory 240).

Ar

10 mm

Fig. 243. *Ficus pumila*. A climbing fig. Juvenile stage with small asymmetrical leaves and adventitious roots, adult stage with large leaves, and fruits. Ar: adventitious roots.

Fig. 244. *Alstonia macrophylla*
The apical meristem of the orthotropic (246) axis has ceased
to grow and its tissue has differentiated into mature
parenchyma cells. Model of Koriba (**295h**, c.f. later stage,
266).

The permanent loss of meristematic activity by a
shoot apex can have as much influence on the
construction of the plant as does the growth
potential of the shoot in the first place. Persistent
development of one apical meristem will result in
monopodial growth (250). Very often, however,
the extension of a shoot will cease because the
apical meristem is lost due to the formation of a
terminal structure such as a flower or
inflorescence. Continued vegetative growth is
then sympodial (250). Alternatively the loss of
meristematic activity may be caused by the
abortion of the apex; the apical meristem dies.
Such death is often not a random happening or
due to damage, but is just as predictable as might
be the commencement of growth of a bud in the
first place (242). Death of the apical meristem in
woody plants often takes place at the end of the
growing season (**245**), cf. Mueller (1988). The
loss of meristematic activity of the shoot apex
may incorporate a dedifferentiation of cell types.
If the cells of the apex become thick-walled and
lignified before abortion (e.g.
sclerenchymatization) then the shoot can
terminate as a spine (124). Alternatively, the
cells may remain alive as parenchymatous cells
but lose their meristematic functions (e.g.
parenchymatization). Although the cells remain
alive, the shoot apex has effectively undergone
abortion. Parenchymatization occurs as a regular
feature during the construction of some trees
(**244**) and more noticeably in the development of
stem tendrils (122), and within the branching
sequences of some inflorescences (**145c**).
Abortion of the apical meristem of an axillary
bud can take place at a very early stage when
the bud primordium is barely formed. No trace of
a bud will then be visible in the axil of the
subsequently developing leaf. In addition a shoot
or root apex may be aborted by outside agency
such as insect damage; the apex may simply be
destroyed, form a gall (278), or in some cases
become parenchymatized. The death and loss of a
whole shoot is referred to as cladoptosis (268).

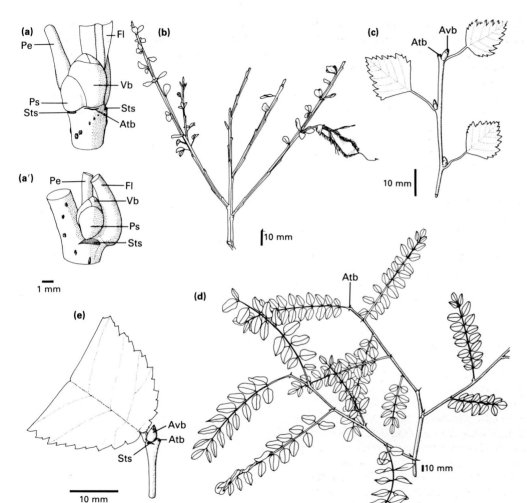

Fig. 245. a) *Tilia cordata*, distal end of shoot with aborted terminal bud. The inflorescence peduncle is in the axil of the foliage leaf, the vegetative bud is in the axil of one of its prophylls, a scale leaf, the other being the inflorescence bract (**235e**). a¹) ditto, lateral node; b) *Cytisus scoparius*, the end of every shoot aborts; c) *Betula pubescens* ssp. *odorata*; d) *Robinia pseudacacia*; e) *Ulmus glabra*, end of aborted shoot. Atb: aborted terminal bud. Avb: axillary vegetative bud. Fl: foliage leaf. Pe: inflorescence peduncle. Ps: prophyll scale leaf. Sts: stipule scar. Vb: vegetative bud in axil of prophyll scale leaf.

The term orthotropic growth implies growth in a vertical direction, in contrast to plagiotropic growth away from the vertical and towards the horizontal. However, in the context of plant construction these terms hold much wider implications. An orthotropic shoot will have a different morphology to that of a plagiotropic shoot of the same species (**246, 247a**). The potential of a shoot expressed in terms of orthotropy or plagiotropy can be a crucial aspect of the form of the whole organism, and is most easily appreciated where these two growth orientations occur on the same plant and do have contrasting morphologies. The bamboo *Phyllostachys* has a plagiotropic underground rhizome system bearing scale leaves. The buds in the axils of these leaves develop mostly into branching orthotropic shoots, with foliage leaves except at the base, and which have a flowering potential (**195d**). Cocoa (*Theobroma*) has two types of branch; the orthotropic chupones which have spiral phyllotaxis, and the plagiotropic jorquette branches which have distichous phyllotaxis. Each type if rooted as a cutting maintains its orthotropic or plagiotropic characteristics (see topophysis 242). In other arborescent plants the morphological distinction of these two branch types may also extend to leaf shape, potential for flowering, reorientation of leaves by internode twisting, and often proleptic growth of orthotropic branches and sylleptic growth of plagiotropic branches (262). It is possible for a single shoot, i.e. the product of a single apical meristem, to change from one form to the other. Thus for most sympodial rhizome systems, the proximal part of each sympodial unit is plagiotropic with one set of morphological features and then turns abruptly upwards, the distal end being orthotropic with a different set of morphological features. Conversely the sympodial units of some lianes (**309e**) have an orthotropic proximal end climbing a support with a plagiotropic distal end growing away from this support.

Within one plant there can be a continuum of branch construction from orthotropism to plagiotropism. This is particularly so where a tree exhibits a metamorphosis (300) of branching types during its life. For example, the distal end of each newly produced plagiotropic branch may be progressively more orthotropically inclined than the previous branch (**301b**). Where a sympodial succession of orthotropic branches develops, the smallest newest distal units may appear to be plagiotropic in a purely orientational sense. This series is referred to as an orthotropic branch complex (**247b**). Similarly a monopodial orthotropic branch may droop secondarily as it elongates but its distal end will always demonstrate its orthotropic origins (**247c**). A plagiotropic sympodial branch system if developed by apposition rather than by substitution (**247d**, 250) may superficially resemble an orthotropic branch complex, but its nature is revealed by the strictly plagiotropic nature of the proximal end of each sympodial unit. A monopodial axis will be referred to as plagiotropic if it is always horizontal in its distal extremity even if the proximal portion resumes a somewhat oblique orientation (**247e**). In these less precise cases, a study of the branch development at different stages will be necessary to identify the plagiotropic or orthotropic tendencies (304).

Fig. 246. *Laetia procera* Leaves and therefore their axillary shoots are spirally arranged ($\frac{3}{8}$ phyllotaxis **221b**) on the orthotropic, vertical axis, whereas the foliage leaves on the plagiotropic axes have a distichous (2-rowed) phyllotaxis. Model of Roux (**291h**).

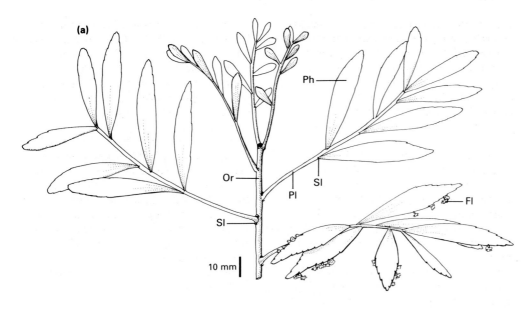

(a)

Ph

Or

Pl

SI

SI

Fl

10 mm

Fig. 247. a) *Phyllanthus angustifolius*, contrasting shoot morphologies; b) orthotropic branch complex; c) orthotropic branch with proximal droop; d) plagiotropic branch sympodial by apposition (250); e) plagiotropic branch with oblique proximal section. Fl: flower. Or: orthotropic shoot bearing plagiotropic shoots only. Ph: phylloclade (126). Pl: plagiotropic shoot bearing phylloclades only. Sl: scale leaf.

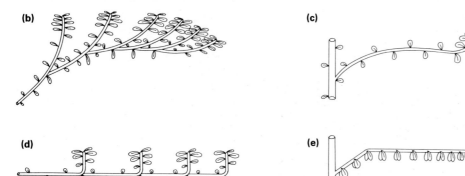

(b)

(c)

(d)

(e)

The potential of an axillary shoot to develop can frequently be considered in terms of its position in proximity to the apical meristem of the shoot bearing it. In this context, potential will be represented by the degree of elongation of the lateral shoot and also the timing of its development. Meristem potential, position, and time of activity should not be considered in isolation. The inhibitory influence of an apical meristem on more proximal axillary meristems is generally called apical dominance. However, this phrase encompasses a range of different and complex phenomena. The main shoot may exert a strong apical dominance on its axillary buds of the current season's growth (**249d**). In the second season these buds may grow rapidly even overtopping the main shoot which is now exerting weak apical dominance (**249e**). Conversely, the axillary bud may develop in the same season as the leader (**249f**) and continue to develop in the following season but always in a subordinate manner to that of the leader (**249g**), a condition known as apical control (Zimmermann and Brown 1971). Apical control implies a more precise influence of an apical meristem on daughter branches. The implied physiological mechanisms behind such controls must be many and varied. Also it is possible that axillary buds because of, or in spite of, imposed apical control have their own built-in 'fate' (topophysis 242) depending upon their position and time of appearance in the branching sequence. The degree to which successive axillary branches along a season's axis elongate has given rise to three terms describing three

Fig. 248a. *Stewartia monodelpha*
Basitonic branching (cf. Fig. **249b**).

contrasting side shoot architectures: acrotony, basitony, and mesotony (**249c**). The distal branches grow more vigorously in the acrotonic condition (**249a**), the proximal branches grow more vigorously in the basitonic condition (**249b**). The same phenomenon can occur in non-seasonal environments if the plant itself has rhythmic growth (exhibited by successive units of extension (**284**).

Fig. 248b. *Fagraea obovata*
Acrotonic branching (cf. Fig. **249a**); the side shoots are overtopping the main shoot.

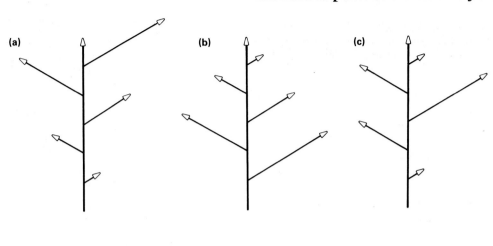

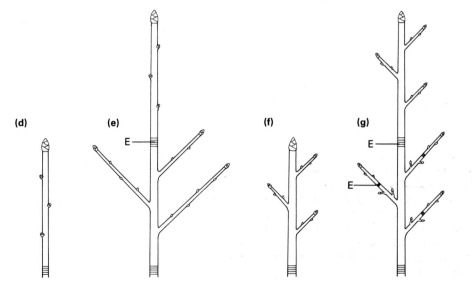

Fig. 249. a) acrotonic development; b) basitonic development; c) mesotonic development; d) strong apical dominance in first season; e) weak apical dominance in second season; f, g) apical control each season. E: end of a season's growth.

Fig. 250. *Cecropia obtusa*
A monopodial trunk. Model of Rauh (**291g**).

The framework of a plant is built up of a number of branches (axes, shoots; these are all somewhat interchangeable and often ambiguous terms 280). A single branch, regardless of age or size, must be constructed in one of two ways. It can be developed by the vegetative extension of one apical meristem (which may rest from time to time as a terminal bud giving rhythmic growth 260) to form a single shoot or shoot unit. The axis thus formed is a monopodium and its structure monopodial (**251f**). Alternatively the axis is built up by a linear series of shoots units, each new distal shoot unit developing from an axillary bud sited on the previous shoot unit. The whole axis then constitutes a sympodium, formed by sympodial growth, and each member of the series derived from one apical meristem is termed a sympodial unit (or caulomer) (**251g**). The sympodial unit plays an important part in an aspect of plant architecture as represented by the module or 'article' (286). A monopodial axis will bear its own axillary shoots each with a fixed or flexible potential for development. The monopodium itself may be indeterminate, i.e. capable of more or less indefinite extension (as for the trunk of a coconut palm *Cocos nucifera*), or it may be determinate in growth, i.e. its apical meristem eventually but inevitably ceases vegetative growth and aborts (244) or differentiates into a non-meristematic structure incapable of continued extension such as an inflorescence (e.g. the trunk of a talipot palm, *Corypha utan* preface). No axillary bud takes over the further extension of this axis. The individual units of a sympodium are likewise either

determinate or indeterminate. In the former case each sympodial unit terminates due to loss of meristematic activity. The apical meristem may die, undergo parenchymatization (244), or develop into a structure such as a tendril (**309a**), spine (**125e**), flower or inflorescence. Each sympodial unit thus formed may be morphologically very similar to the last ('article' 286). A sympodium developing in this manner is said to be sympodial by substitution (**251d**). If each sympodial unit is indeterminate, it will continue its apical vegetative growth in a subordinate manner usually deflected at an angle from the line of growth of the sympodial axis. Nevertheless it can then continue to grow indefinitely, bearing its own axillary shoots, often with a reproductive potential. This type of axis is sympodial by apposition (**251e**).

(Continued on page 252.)

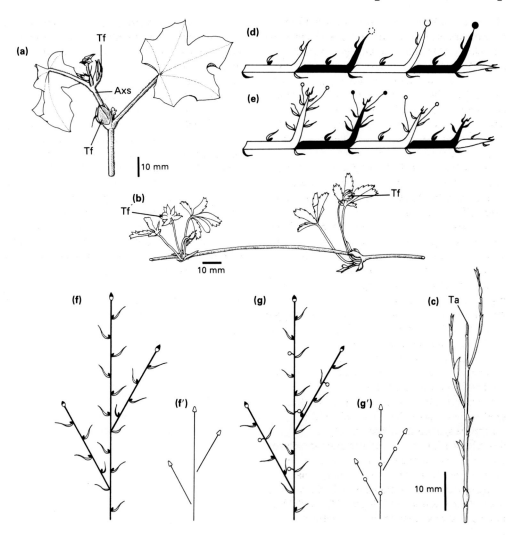

Fig. 251. a) *Fremontodendron californica*, end of sympodial shoot (see **252** and **297f**); b) *Potentilla reptans*, sympodial runner (134); c) *Cytisus scoparius*, sympodial growth due to abortion of apex (244); d) sympodial growth by substitution, alternate sympodial units in black; e) sympodial growth by apposition; f) monopodial growth; f¹) shoots units present in 'f'; g) sympodial growth, each shoot unit terminating in a flower (circle); g¹) shoot units present in 'g'. Axs: axillary shoot. Ta: terminal abortion. Tf: terminal flower.

Fig. 252. *Fremontodendron californica*
Sympodial growth, a flower terminating each sympodial unit, growth of the trunk being continued by an axillary bud (Figs. **251a** and **297f**).

The sympodial or monopodial nature of an axis (250) may be clearly apparent, or may only be deduced by careful scrutiny of relative bud and leaf positions (e.g. Vitidaceae 122, *Carex* spp. 234, *Philodendron* sp. 10). A sympodial axis often superficially looks like a monopodial axis with lateral shoots developing from it (**251g**). The origin of a sympodial system of branching can be further disguised by secondary thickening of the axis, the proximal portion of each sympodial unit enlarging in girth and the free ends of each unit remaining unthickened (**304**). In a sympodial system, more than one bud may develop to replace the 'lost' parent shoot apex. If two such renewal shoots develop in close proximity then a 'Y'-shaped branching will occur giving an appearance of dichotomy (**130**) (see false and true dichotomy 258). If sympodial renewal shoots are produced singly, the axis as a whole cannot branch and such shoots have been termed regenerative (Tomlinson 1974). If more than one renewal shoot is formed, branching must occur and is said to be proliferative. Plants may be constructed of either monopodial or sympodial shoots, or an organized mixture of the two. Most rhizomatous plants are sympodial. The traditional descriptions of inflorescence architecture (140) are based on a distinction between monopodial (raceme) or sympodial (cyme) construction. If an axis is determinate because it is terminated by flowering it is said to be hapaxanthic. Conversely, a shoot with lateral flowering and therefore potentially monopodial growth is said to be pleonanthic (cf. terminology used for sympodial orchids 253a, c, d).

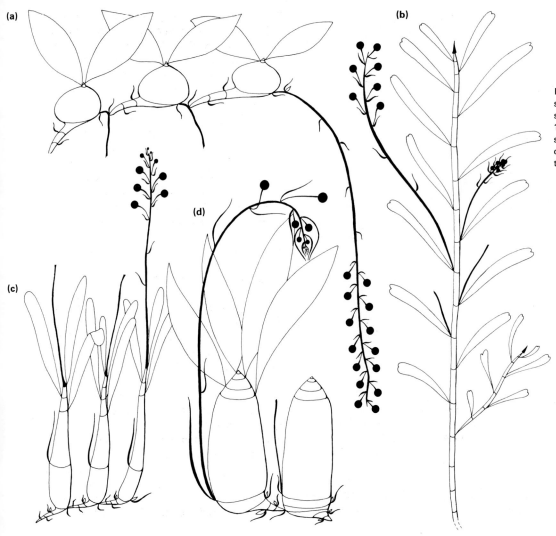

(a)

(b)

(c)

(d)

Fig. 253. Four contrasting growth forms of orchids. a) sympodial with vegetative shoot unit borne on reproductive shoot unit (*Gongora quinquenervis* type, after Barthélémy 1987); b) monopodial with lateral inflorescences; c) sympodial with terminal inflorescences (acranthous type); d) sympodial with lateral inflorescences (pleuranthous type).

The application of the terms short shoot and long shoot is to a large extent self-explanatory. In many perennial plants, particular if woody, some shoots have relatively long internodes and thus leaves which are well spaced from each other. These long shoots are frequently described as having an exploratory capacity, extending the framework of the plant into new territory. Other shoots on the same plant, in contrast, may extend very little in each growing season having very short and usually few internodes. These short shoots are then said to have an exploitary capacity, producing leaves at the same location season after season. The longevity of both long and short shoots varies with the species, the environmental location of the plant, and the position of the shoot within the plant's framework. Both long and short shoots may abscise (268) after a few seasons or persist more or less indefinitely. The location of a new axillary bud developing in such a system can influence its potential to extend as either a long or short shoot. For example, proximal buds on a long shoot may have a potential to form short shoots whilst distal buds will form long shoots (i.e. in this example acrotonic development **249a**). The potential of a bud may be apparent from the number and type of unexpanded leaves it contains, or potentially long or short shoot buds may be indistinguishable. A predictably short shoot bud may occasionally develop as a long shoot if its bearing axis is damaged. Both types of shoot, but especially short shoots, often have a precise and consistent number of foliar components for each increment of growth. These

Fig. 254. *Acer hersii*
Determinate lateral short shoot systems borne on an indeterminate long shoot. This combination of 'exploratory' and 'exploitary' branching occurs in many unrelated plants, cf. diagram of bamboo and ginger rhizome systems (Fig. **311**).

details for any one species can be quite elaborate as, for example, in *Cercidiphyllum* (Titman and Wetmore 1955). Short shoots are often involved in flowering or in the production of spines (124). Strictly speaking a spine or a flower represents a short shoot in itself. Epicormic branching and cauliferous branching (240) also represent forms of short shoot formation. Both long and short shoots can be either monopodial or sympodial in their construction, and in many plants each can change its potential and switch from one type to the other in certain circumstances (**255c**). Short shoots usually develop initially as axillary buds on existing long or short shoots (lateral short shoot); however the distal portion of a sympodial unit in a sympodium formed by apposition (**251e, 304**) can also take on the form of a (terminal) short shoot.

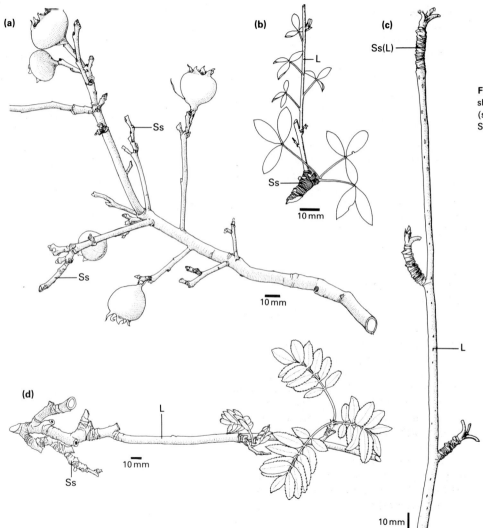

(a)

Ss

Ss

Ss

10 mm

(b)

L

Ss

10 mm

(c)

Ss(L)

L

10 mm

(d)

L

Ss

10 mm

Fig. 255. Examples of the intermixing of long and short shoots. a) *Mespilus germanica*, b) + *Laburnocytisus adamii* (see **274**), c, d) *Sorbus* spp. L: long shoot. Ss: short shoot. Ss(L): long shoot changed to short shoot.

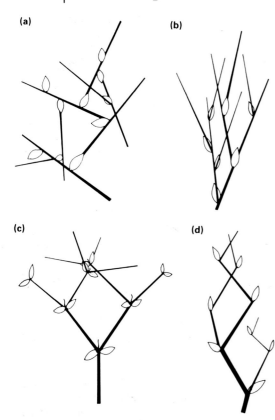

(a)

(b)

(c)

(d)

Fig. 256. Diagram of types of divarication: a) wide angle, b) fastigiate, c) sympodial branching, d) zigzag.

The term divarication or divaricating shrub is applied to a growth form that is distinctive, and in its narrowest sense refers to a densely branched plant endemic to New Zealand and in its widest sense to any similar shrub of usually dry windswept habitats. These latter types are typically spiny; the New Zealand divaricates are distinctive in that they are seldom spiny and occupy sheltered woody areas as well as exposed ones (Tomlinson 1978). Divaricate implies 'wide-angled' and this roughly describes the branching of these shrubs, cf. **256b**, which is such that a three-dimensional interlacing of twigs results. The twigs are usually fine, wiry, and tough. A common observation is that any one branch, if severed, cannot easily be untangled from the remainder. Such apparent chaos is in marked contrast to the orderly patterns of branching frequently met with (228). The divaricate framework develops in a number of ways or combinations thereof. Lateral branches may be produced at a wide angle at, or exceeding, 90° (**256a**); alternatively an equivalent tangle may result from fastigiate branching at narrow angles (**256b**). Sympodial growth can result in a zigzag axis (**256c**) or in some species a monopodial shoot bends at each node to produce a zigzag axis (**256d**). New Zealand divaricate shrubs share a number of distinctive features although they are not necessarily all present in any one species:

(1) abrupt bends in branches;
(2) often lateral rather than terminal flowering;
(3) thin wiry branches;
(4) lack of spines (although stumps of dead twigs may persist);
(5) three-dimensional interlacing of branches;
(6) simple and small leaves;
(7) leaves inside the tangle larger than those at the periphery;
(8) persistent short shoots (254); and
(9) accessory buds (236).

A spiny divaricate shrub is relatively protected from large mammalian herbivores; these are absent from New Zealand but it has been suggested that the extinct Moa, a large flightless herbivorous bird, would have been deterred by divarication, but not by spines. Another form of contorted growth occurs as a 'sport' in some shrubs (280).

Fig. 257. a) *Sophora tetraptera*, zigzag divarication (**256d**); b) *Corylus avellana*, divaricating type growth in a contorted mutant form; c) *Rubus australis*, apparent divaricating type growth, in fact largely due to leaf form, cf. **77c**; d) *Bowiea volubilis*, a much branched divaricate climbing inflorescence (144).

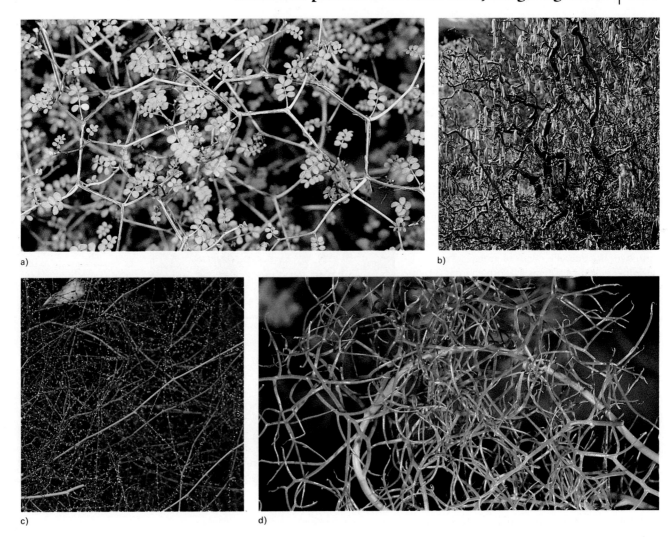

a)

b)

c)

d)

Dichotomy implies the bifurcation of an axis into two more or less equal halves, a fork. With very few exceptions when this occurs in the branching of a flowering plant, the two arms of the fork have developed from axillary buds situated close behind the distal end of the parent axis, the apical meristem of which has ceased to function. The apical meristem may have aborted (244) or formed a temporary structure such as an inflorescence. If this sequence is repeated a regularly branched sympodial pattern can develop (130). This type of bifurcation is called false or pseudo-dichotomy. Dichotomy (true dichotomy) is confined to development in which an apical meristem, without ceasing cell division or any partial loss of activity, becomes organized into two, not one, directions of growth. Two terminal apical meristems are formed from the one. True dichotomy has been recorded in *Mammillaria*, *Asclepias* (dicotyledons), and in a number of monocotyledons—*Chamaedorea*, *Flagellaria*, *Hyphaene* (295a), *Nypa*, and *Strelitzia* (258) and in *Zea* as a teratology (270) (Mouli 1970), seedlings having twin epicotyls. Careful anatomical study of the developmental sequence is necessary in order to identify the occurrence of a true dichotomy. False dichotomy resulting from sympodial growth (259c) will closely resemble a true dichotomy (259b) especially if all trace of the parent shoot apex is lost. A similar condition can arise if an axillary shoot develops very precociously at the apical meristem (259d). In all cases, the location of leaves and prophylls will give an indication of which arrangement is present, but mirror imagery of the branch pair does not necessarily indicate true dichotomy which can occur with (259b) or without (259a) mirror imagery. The ground plan diagram for a true dichotomy can however be very similar to that of a false dichotomy and drawing up such diagrams becomes an exercise in 'spot the difference'.

Fig. 258a, b. *Strelitzia regina*
Two stages in the development of a young plant showing the simultaneous production of two leaves from initially one apical meristem indicating that it has dichotomized.

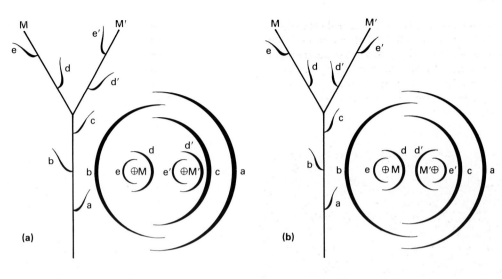

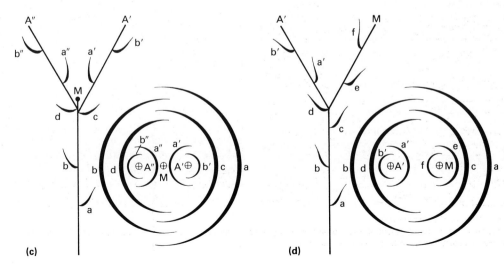

Fig. 259. Pairs of stick and ground plan diagrams representing true and false dichotomy. Corresponding leaves are appropriately labelled in each pair. a) True dichotomy without mirror imagery; b) true dichotomy with mirror imagery; c) false dichotomy resulting from sympodial growth; d) false dichotomy resulting from precocious branch development. A: axillary shoot. M: main shoot.

Fig. 260a. *Terminalia catappa*
Rhythmic growth leading to tiers of branches at the location
of each pseudowhorl (218) of leaves on the orthotropic axis
(246). Model of Aubreville (**293d**).

Fig. 260b. *Phyllanthus grandifolius*
Continuous growth with evenly spaced branches.

In a uniform climate or environment it is possible
for a plant to grow continuously with constant
production of leaves and axillary shoots. Such
axillary meristems may have no dormancy and
no protected resting phase (sylleptic growth 262).
In a climate with clearly defined seasons
alternating between periods favourable and
unfavourable for growth, shoot extension is likely
to be rhythmic. Expansion growth will cease for
the duration of the dry or cold season during
which time apical meristems may be protected in
some manner (264). Some plants grow
continuously even in a seasonal climate. *Carex
arenaria* (**235a**) rhizomes continue to extend in
the cold season although internodes produced are
shorter than those developing at other times.
Conversely many plants grow rhythmically even
in a completely even environment, and then
individuals of a species are likely to be out of
synchrony with each other, or even different
parts of the same plant may be undergoing
different phases of leaf production or flower
production (manifold growth). Shoots of tea
(*Camellia sinensis*) extend in a rhythmic manner,
alternating between foliage leaf and cataphyll
(64) production several times a calendar year
regardless of climate or seasonal change. Lack of
apparent growth during rhythmic development
does not necessarily indicate a cessation of
meristematic activity. At a time when no
outward sign of growth is to be seen, intensive
cell division and differentiation of organs may be
taking place in the apparently dormant apical
meristem. This time is referred to as the time of
morphogenesis and its location the unit of

morphogenesis (284). Subsequently the distal part of the shoot will undergo readily visible extension, mostly due to cell enlargement, this portion of the stem being referred to as the unit of extension (Hallé and Martin 1968). Periods of morphogenesis and extension follow in sequence and may overlap (**283i**). The location of a temporary cessation of extension is usually, but not necessarily, indicated by the presence of crowded internodes and scale leaf scars. An alternation of large and small leaf scars or other features can occur on a shoot growing continuously due to production of different organs in regular sequences. Nevertheless rhythmic growth is likely to result in the clustering of axillary meristems having equivalent potentials along the shoot, as in acrotonic or basitonic growth (248). Thus pseudowhorls of branches often result on axes having rhythmic growth (**260a**); branches on axes having continuous growth are more likely to be located at regular intervals (**260b**). This distinction in branch formation is one criterion incorporated in the description of architectural models of tree form (288).

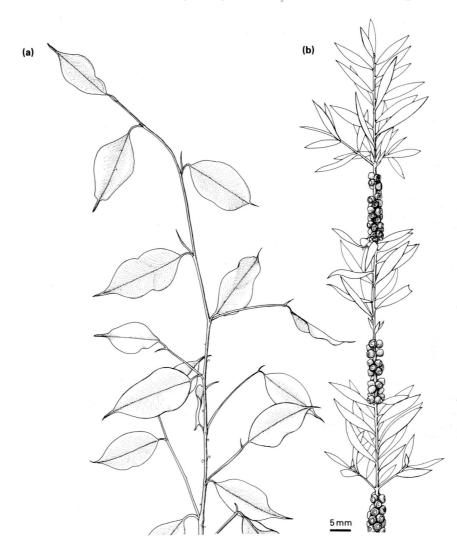

(a)

(b)

5 mm

Fig. 261. a) *Ficus benjamina*, continuous growth; b) *Callistemon citrinus*, rhythmic growth, rest periods marked by transition between vegetative and reproductive development (indicated by persistent fruits).

As the apical meristem of a shoot develops (stem development 112) it produces a succession of leaf primordia. Situated in the axil of each leaf primordium is a newly formed shoot primordium which represents the apical meristem of a potential axillary shoot. Each axillary shoot primordium will have one of two immediate fates other than abortion (244). It may become organized into a temporarily dormant protected resting structure, a bud, or it may develop and grow without delay and thus extend contemporaneously with the apical meristem of the plant shoot. Such growth of a lateral shoot without a resting stage is referred to as sylleptic growth giving rise to a sylleptic shoot. Growth of a dormant bud is referred to as proleptic growth, forming a proleptic shoot. The distinction is based on the time of extension of the axillary shoot primordium and the event is usually recorded morphologically at the proximal end of an axillary shoot by the presence or absence of protective structures or their scars associated with the presence (prolepsis) or absence (syllepsis) of a protected resting stage. As a sylleptic shoot has grown without delay, the first leaf or leaves (prophylls 66) are borne some way along its axis; the portion of stem proximal to the prophylls being termed the hypopodium (**263d**). Thus the presence of a hypopodium usually indicates sylleptic growth although such a structure can be found in plants that have developed proleptically with naked buds (bud protection 264). Conversely some sylleptic shoots have very short hypopodia and transition of leaf types disguising the sylleptic origins. Sylleptic

Fig. 262. *Persea americana*
Sylleptic growth with no resting phase and no bud formation. A long interval of stem (hypopodium) separates the first pair of leaves (prophylls 66) on the side shoot from its parent axis.

growth is usually associated with a tropical environment and a given plant species will frequently bear both sylleptic and proleptic shoots. Often a leaf will subtend two axillary meristems (accessory buds 236), one developing sylleptically and one remaining dormant as a bud representing a potentially proleptic shoot. These two shoot types will typically have different potentials within the branching architecture of a plant. In trees, proleptic branches are often orthotropic whereas sylleptic branches are often plagiotropic (246).

Etymology
Syllepsis: taking (happening) together, i.e. terminal and axillary shoot extension is simultaneous.
Prolepsis: an anticipated event—extension from an initially dormant bud (originally applied to precocious growth of a dormant bud expected to rest through an unfavourable season).

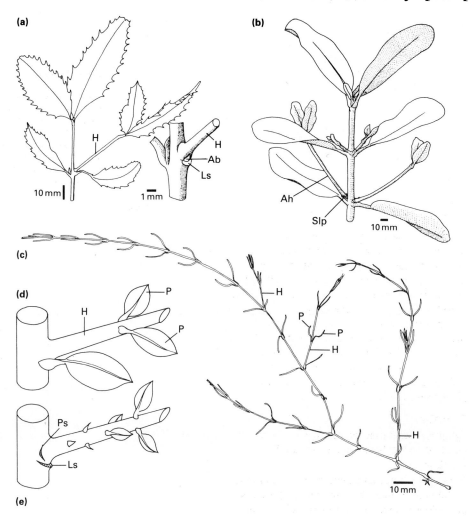

(a)

H

10 mm

H
Ab
Ls

1 mm

(b)

Ah

Slp

10 mm

(c)

H

P
P
H

H

10 mm

(d)

P
H
P

(e)

Ps

Ls

Fig. 263. a) *Doryphora sassafrass*, sylleptic; b) *Clusia* sp., proleptic in this specimen, axillary shoot initially dormant protected by prophylls; c) *Gypsophila* sp., sylleptic; d) sylleptic node; e) proleptic node. Ab: accessory bud (236). Ah: apparent hypopodium, in fact second internode of axillary shoot. H: hypopodium. Ls: leaf scar. P: prophyll. Ps: prophyll scar. Slp: scale leaf prophyll formerly protecting axillary bud. d, e) Redrawn from Tomlinson (1983).

Whilst in a short- or long-term dormant state, the vulnerable meristematic tissue of a shoot apex is usually protected from cold or desiccation and to some extent insect attack by being enclosed in a structure termed a bud. A variety of organs may be incorporated into a bud. For example the protective component may be composed of one to many enveloping scale leaves (**265a**), the single prophyll of a monocotyledon (**66a**), or a pair of prophylls in a dicotyledon (**66b**). When the shoot protected by the bud emerges the bud scale leaves may be seen to form a heteroblastic series (28). Bud scales often represent only the base of a leaf, the petiole and lamina not being developed (**29d**). Stipules attached to the leaf may assist in bud protection (**265e, 52, 55o**) or indeed the entire bud may be built up of one or more stipules (**265c**). Hairs, frequently glandular, may be incorporated into the bud construction. Elaborate glandular hairs in this context are called colleters (80). Axillary buds and sometimes terminal buds (**265b**) are often protected to a greater or lesser extent by the enveloping base of the subtending leaf (**265f, 51c**) before it is shed. This is almost inevitably the situation in monocotyledons where the leaf is inserted around most or all of the stem circumference. Such protective leaf bases may persist after the distal part of the leaf has been abscissed. Bud protection may result from the secretion of gums and waxes (**264a**) or by the elaboration and lignification of associated stem tissue forming a woody construction above the shoot apex (**264b**). A naked bud is one in which the shoot has temporarily ceased to grow, the most recently formed young leaves themselves forming the bud; these leaves will not be shed when the shoot recommences growth, but will expand to form fully sized foliage leaves (**265d**). Flower buds are similarly protected by bracts (62), stipules of bracts, and by the more proximal perianth segments themselves.

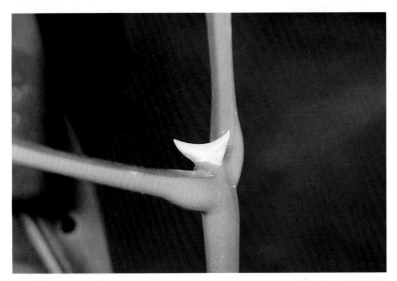

Fig. 264a. *Potalia amara*
Apical meristem protected by a distinctively shaped excretion of wax.

Fig. 264b. *Palicourea* sp.
Apical meristem protected within a dome of parenchymatous tissue.

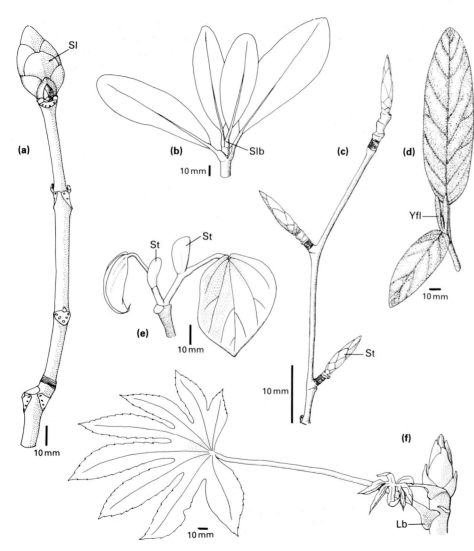

Fig. 265. a) *Aesculus hippocastanum*, terminal bud of scale leaves; b) *Clusia* sp., shoot apex hidden by pair of leaf bases; c) *Fagus sylvatica* (cf. **119g**), bud composed of stipules; d) *Viburnum rhytidophyllum*, shoot apex hidden by young leaves; e) *Exbucklandia populnea*, shoot apex hidden by stipules; f) *Fatsia japonica* (cf. **51c**), bud hidden by leaf base. Lb: leaf base. Sl: scale leaf. Slb: swollen leaf base. St: stipule. Yfl: young foliage leaf.

Fig. 266. *Alstonia macrophylla*
An initially laterally growing branch has bent at its base and is reoriented into a vertical position, top of picture. Sister lateral branches remain horizontal (cf. earlier stage **244**). Model of Koriba (**295h**).

Generally speaking a plant organ will grow in a particular orientation with respect to gravity and light. Other factors may be involved, many horizontally growing underground rhizomes can extend along at an adjustable distance beneath an uneven soil surface. Changes in the orientation of an organ such as a shoot or branch of a tree, fall into five distinct categories and may be brought about in response to either external or internal factors.

(1) A stem may incorporate one or more pulvinus (128) and can bend at this point.

(2) A branch may become reorientated regaining a lost position by unequal cambial activity on opposite sides of the axis.

(3) A branch may become progressively bent or arched as a consequence of its own weight and lack of self support (e.g. model of Champagnat **293b**).

(4) An axis may begin to grow in a new direction due to a change in the directionality of environmental factors.

(5) The orthotropic or plagiotropic potential of the axis may change (246, 300).

In the context of organized plant construction, two mechanisms occur as part of the innate dynamic morphology of a plant. Change in direction of growth (as in 5 above) is probably a frequent phenomenon. It usually includes a progression from plagiotropic growth to orthotropic growth (246) or vice versa and may be accompanied by a complete change in the morphological features of the axis

(metamorphosis 300). The second form of innate change of orientation involves the change in dimension and shapes of existing cells resulting in a bending or repositioning of the existing structure (a similar event occurs in contractile roots 106). Bending of an existing axis occurs as a matter of course during the development of many trees (Fisher and Stevenson 1981) and is a diagnostic feature of some of the models of tree architecture (288). One branch only of a whorl of horizontal branches of some species of *Alstonia* (**244, 266**) (model of Koriba **295h**) will bend at its base to become a vertical component of the trunk. A more gradual reorientation occurs in the trunk of some trees (model of Troll), the distal growing extremity of which is always arched over but which subsequently becomes vertical (**293g**). Conversely the seedling axis of *Salix repens* grows vertically before bending at its base to become prostrate (**267b**). A secondary change in orientation occurs in the inflorescence or flower axis of some plants resulting in the deposition of fruits into water, soil, or rock cavities (**267a, c**).

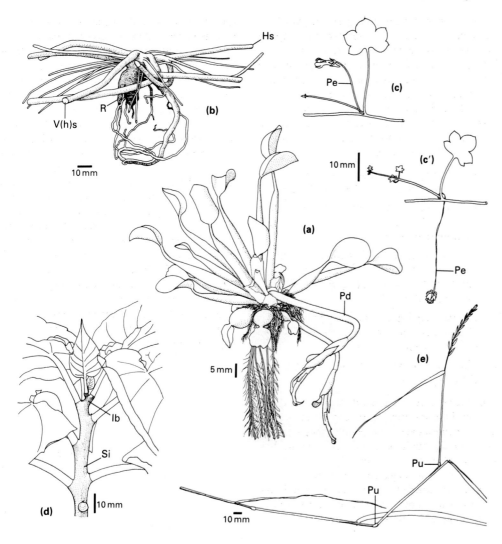

Fig. 267. a) *Eichhornia crassipes*, rosette of floating leaves; b) *Salix repens*, young plant with bent seedling axis; c) *Cybalaria muralis*, flowering node; c¹) ditto, fruiting node; d) *Cyphomandra betacea*, straightening internodes; e) *Agropyron* (*Elymus*) *repens*, stem bent at node by enveloping leaf pulvinus. Hs: horizontal side shoot. Ib: initially bent internode. Pd: peduncle, secondarily lowered and bent. Pe: pedicel. Pu: leaf pulvinus. R: root. Si: straightened internode. V(h)s: initially vertical seedling axis, now horizontal.

Fig. 268a. *Acacia dealbata*
Loss of distal leaflets in bipinnate leaf and loss of distal half of remaining leaflets.

Fig. 268b. *Phyllanthus grandifolius*
Cladoptosis of phyllomorphic branches (i.e. temporary branches resembling compound leaves). Model of Cook (**291e**).

Cladoptosis is the term applied to the fall or shedding of branches. Many plant parts are jettisoned in a positive manner usually by the formation of some sort of abscission zone of dying cells which isolate the organ. Stipules, leaves (or parts of leaves **268a**), flowers, inflorescences, fruits, and seeds, all may be shed in this manner. Vegetative shoot structures are also shed particularly from perennial plants. These detached structures are either living propagating units (vegetative multiplication 170) or structures that have died or die as a result of the process of shedding. The latter include organs or parts of organs that die soon after their initiation (abortion 244). Cladoptosis is specifically applied to the process of shedding of the whole or part of single branches or complexes of branches which normally will have been developing successfully for some period. In some instances the branch will have died down to its point of attachment to another branch and will then rot or break at this point. In many *Eucalyptus* species the stump of the dead branch remains embedded in the trunk of the tree which will encroach by growth around it. Only later does the proximal part of the stump become detached within the trunk, the process being accompanied by the production of gum and the remains of the branch is shed. In other plants an abscission zone develops at the point of attachment of a live branch which is thus isolated, this process of cladoptosis resembling the shedding or articulation (48) of leaves of deciduous plants (**269a–c**). The shed branch may be many seasons old and of considerable size. Aerial shoots which represent the distal portion of underground sympodial rhizome axes are similarly shed either by death followed by rotting or by the formation of an abscission zone (**269d**). The loss of components of the branching system in a plant is in many ways just as important and apparently organized as the controlled growth of these components in the first place (**280**). The branches of some trees resemble large compound leaves and are called phyllomorphic branches (**268b**). They are shed in a similar manner to leaves, thus becoming 'throw away' branches (model of Cook **291e**).

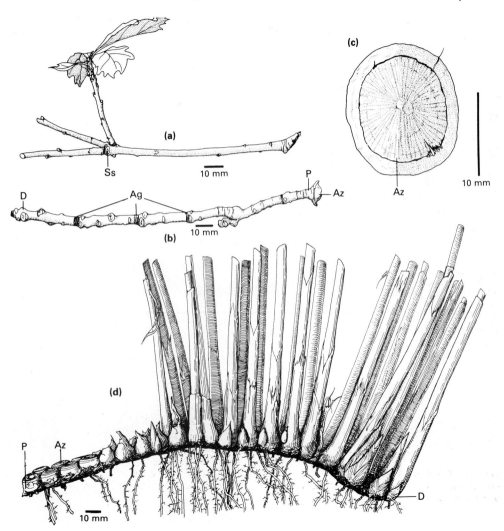

Fig. 269. a) *Quercus petraea*, shoot with scar of shed lateral shoot; b) ditto, jettisoned old shoot; c) proximal end of 'b'; d) *Cyperus alternifolius*, rhizome, successive sympodial units (250) alternate left and right. Az: abscission zone. D: distal end. P: proximal end. Ss: shoot scar.

Teratology refers literally to the study of monsters. The production by a plant of a structure that is atypical of its normal morphology is thus described as teratological; what constitutes 'normal' morphology for a plant is not necessarily easy to decide. Many plants have the ability to produce unusual structures in response to unusual environmental phenomena. The proliferation (176) of grass inflorescences particularly in response to damp conditions is not necessarily teratological but rather the normal behaviour of the plant in unusual conditions. Teratological malformations may be induced by an internal genetic disruption or by interference of a developmental sequence by a substantial change in environmental factors (cold, drought), by animal activity, or by chemical factors. Pest and weed controlling chemicals frequently promote a teratological response. More common morphological 'mistakes' include the development of galls (278), and fasciation (272). Sometimes the adnation of parts represents a teratological event (230a) or it may be the usual mode of development of a plant (231b). Other noticeable forms of teratological events affect the shape of leaves, such as peltation (271f), the production of leaf-like structures in place of perianth segments in a flower (270), and the development of actinomorphic flowers in a plant that typically has zygomorphic flowers (peloric development). Another form of teratological malfunction is concerned directly with constructional organization and bud potential (topophysis 242). Within the branching framework of the plant certain buds in particular

locations often have a fairly predictable fate, to develop into an inflorescence or to develop into either a long shoot or a short shoot, for example. When mistakes occur within these relatively inflexible organizations, buds may develop into the 'wrong' structure or in the 'wrong' position or at the 'wrong' time. Two examples are shown for *Solanum tuberosum* (**271g, g¹**). Groenendael (1985) lists the range of teratological constructional sequences found in *Plantago lanceolata*. The plant has a set of internode types (metamers 282), e.g. long internodes, short internodes, nodes with foliage leaves, nodes with bracts, nodes with flowers. Only one sequence

gives a normal plant (**145e**). Departures from that sequence give distorted morphologies having accurately formed organs in the wrong place (**145e¹**). Most live plant cells are totipotent and teratological activity includes either a confusion of the controlling factors of cell division or the activation of the correct developmental sequences in the wrong place at the wrong time.

Fig. 270. ***Fuchsia*** cv. Mrs. Popple
A flower in which one sepal has developed with the morphology and pigmentation of a foliage leaf.

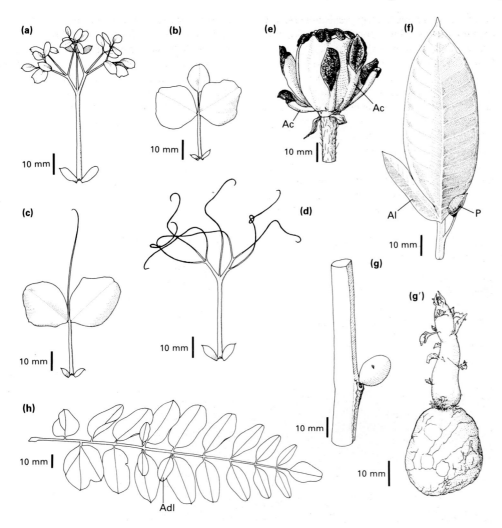

(a)

(b)

(c)

(d)

(e)

Ac

Ac

(f)

Al

P

(g)

(g′)

(h)

Adl

Fig. 271. a–d) *Pisum sativum*, mutant leaf forms (cf. **57e**) (a, P1201; b, P1200; c, P1196; d, P1198; see Young 1983). e) *Papaver orientale*, abnormal fruit (poricidal capsule, **157v**); f) *Plumeria rubra*, single leaf; g) *Solanum tuberosum*, stem tuber on aerial shoot; g¹) ditto, stem tuber on stem tuber (cf. **139e**); h) *Robinia pseudacacia*, single leaf. Ac: additional carpels. Adl: additional leaflet. Al: abnormal lobe. P: peltation (88) of abnormal lobe.

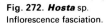
Fig. 272. *Hosta* sp.
Inflorescence fasciation.

A fasciated stem or root (usually seen in adventitious roots 98) is one which is abnormally flattened and ribbon-like. The term is also applied to stems that develop abnormally into a hollow tube (ring fasciation) or with a number of flattened radiating wing-like extensions (stellate fasciation). Many plants produce unusual stem shapes (120) in the normal course of development and are not then fasciated. A fasciation is a teratological phenomenon and may be caused by numerous agencies (270). The developmental nature of any one example can probably only be recognized by careful study. If the flattened root or shoot system (vegetative or reproductive) has come about by the lack of separation of normally distinct organs, then strictly it represents a type of connation (i.e. the joining of like parts 234) and the distal end of the shoot/root apex is composed of a number of laterally adjacent apical meristems. A true fasciation represents the product of a single apical meristem that has become broad and flat instead of a normal dome shape. An abnormally occurring dichotomy (258) will form an apparently fasciated stem if the resulting daughter stems are connate, particularly if the development of the dichotomy is repeated. Flattened stems (cladodes and phylloclades 126) are normal features of some plants, the flat shape developing due to meristematic activity on the flanks of the stem rather than being due to a disruption of the apical meristem.

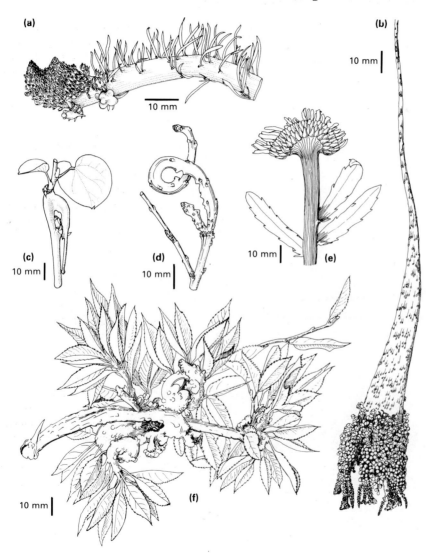

(a)

(b)

10 mm

(c)

10 mm

(d)

10 mm

10 mm

(e)

(f)

10 mm

Fig. 273. Examples of fasciation. a, b, e) flattened inflorescences. c, d, f) distorted vegetative shoots. a) *Linaria purpurea*, b) *Trichostigma* sp., c) *Circis siliquastrum*, d) *Forsythia intermedia*, e) *Chrysanthemum maximum*, f) *Prunus autumnalis*.

A chimera is a structure or tissue that is composed of a mixture of cells from two different sources and therefore of two distinct genotypes. This can come about either by the grafting together of two different species or by mutation within a single growing plant. In the former situation, usually a superficial layer of cells derived from one species is found overlying that derived from a second species (a periclinal chimera). The chimera then develops a morphology that may resemble either donor or be a flexible mixture of the two. A classic example is that of + *Laburnocytisus adamii*, a chimeral tree formed from a graft of *Cytisus purpureus* on *Laburnum anagyroides*. The plant will consist of underlying *Laburnum* cells with a surface layer of *Cytisus* cells. If the *Laburnum* cells break through, a *Laburnum* branch will be formed; if the *Cytisus* cells proliferate at the surface a *Cytisus* branch can be formed. More often branches show a mixture of morphological features derived from both parents (**274, 255b**). Chimeras can also occur naturally if genetic information contained in one cell is altered by whatever cause. All the progeny of this cell will then contain the same altered information and a sector of the plant may have different characteristics from the remaining normal tissue. Factors involved could affect colour, texture, hairiness, or shape. If the mutation occurs in a superficial cell of the plant, particularly at the shoot apex, again a superficial layer of cells of one type may overlie the bulk of cells of the unaltered type. Many plants with variegated leaves represent chimeras with cells of one colour type more or less covering cells of the second colour type. A sectorial chimera is one in which the two different cell populations lie side by side rather than one enveloping the other (**275**) (Tilney-Bassett 1986).

Fig. 274. + ***Laburnocytisus adamii***
A chimera in which either *Laburnum* tissue (large leaves) or *Cytisus* tissue (small leaves) predominates in a haphazard manner.

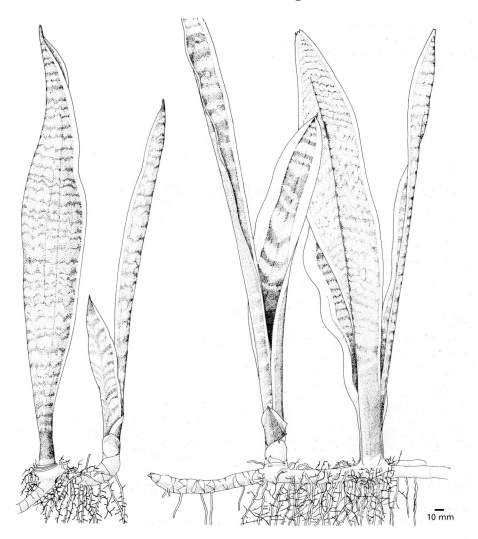

Fig. 275. *Sansevieria trifasciata,* cv. *laurentii.* The edges of each leaf lack pigment and have a different cell parentage to that of the pigmented areas.

10 mm

Small swellings on both leaves and roots may be due to a number of factors. Galls (278) are mostly developed in response to insect attack, and are not therefore a normal morphological feature of the plant. Some swollen cavities are inhabited by ants, mites, and other fauna (domatia 204). In addition a particular range of structures, referred to as nodules, occur as normal features on a limited number of plant species and are inhabited by bacteria which may or may not be presumed to have some symbiotic role. The most common type of nodule is the root nodule typical of many members of the Leguminosae. These nodules contain nitrogen-fixing bacteria and vary in shape ranging from spherical to variously branched (277). They can be very similar in appearance to mycorrhizae (see below). Leaf nodules containing bacteria are of two types. They may occur as small barely noticeable protuberances on the leaf petiole, mid-rib, or lamina, as found in some members of the Rubiaceae (204b) or as a row of small swellings along the leaf edge as in members of the Myrsinaceae (204a). Leaf and root nodules are thus constant features of the plant species on which they occur and represent normal morphological structures developed by the plant in response to normal infection. Nodules formed on roots in association with bacteria are relatively large and distinctive structures. Permanent associations also occur between the roots of very many flowering plants and fungi. This association leads to structural features termed mycorrhizae; the term is sometimes applied to the association itself. Different types of

mycorrhizae have different physiological significance for the participating organisms and may involve distinctly different fungal groups. Only one type of mycorrhiza, the ectomycorrhiza, is likely to be noticed because of some morphological feature of the flowering plant roots, the roots becoming variously branched and ensheathed in fungal tissue (mycelium) (276). However, this is also the case for some plants having an arbutoid mycorrhiza (Harley and

Smith 1983). Other types of mycorrhizae, with little or no externally obvious features are vesicular arbuscular, ericoid, monotropoid, and orchid mycorrhizae. The visible features of an ectomycorrhiza are fungal in origin and do not represent plant tissue developing in response to the presence of the fungus, in contrast to a nodule or gall (278), which is developed by the plant.

Fig. 276. *Fagus sylvatica* Ectomycorrhizal roots (white). Unaffected roots are brown.

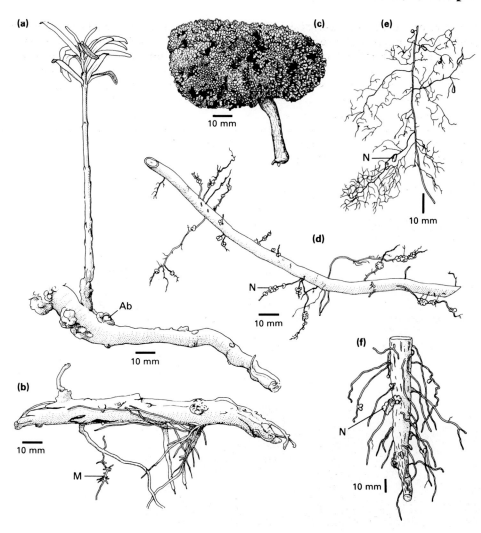

Fig. 277. a, b) *Hippophae rhamnoides*, portions of root bearing adventitious buds and mycorrhizae; c) *Alnus glutinosa*, single massive nodule; d) *Alnus glutinosa*, nodules on minor roots; e) *Acacia pravissima*, single small nodules; f) *Vicia faba*, nodules on minor roots. Ab: adventitious bud (178). M: mycorrhizae. N: nodule.

Fig. 278. _Rosa canina_
The plant tissue has been induced to form a novel structure (the gall) whilst the production of emergences (76, 116) has not been suppressed.

The normal range of morphological features exhibited by a plant may be disrupted or modified in many ways (see teratology 270). One distinctive form of morphological disturbance occurs in response to occupation by a range of fauna including nematodes, mites, and insects, and leads to the development of a gall. A gall is constructed of plant cells and depending upon the organism involved may have an apparently totally disorganized development, or may be a recognizable but distorted morphological feature of the plant concerned. Alternatively it may represent an organized developmental construction that is normally only produced by the plant if stimulated to do so by the animal (**278**). These distinctive structures occur in a wide range of shapes, each shape being typical of attack on one plant species by one particular gall-former. The illustrations here (**279**) are all galls of two oak species (_Quercus petraea_, _Q. robur_), each gall being inhabited by one or more developing animals. One distinctive type of gall, the witches' broom, occurs on a number of tree species and is caused by fungal attack. The tree's response is the over-production of shoots upon shoots, these persisting for several years. Similar witches' brooms can develop in response to mechanical damage.

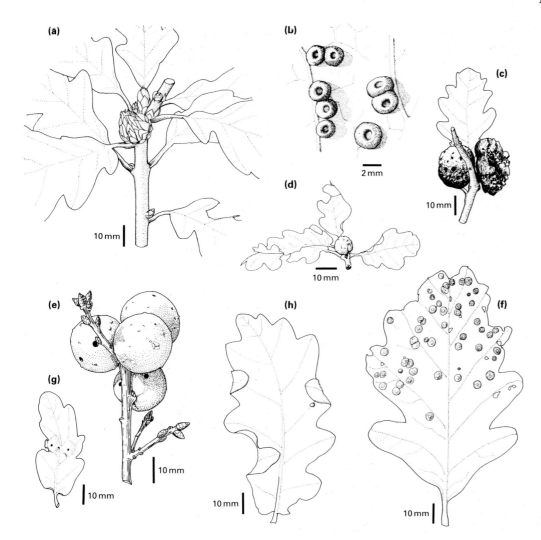

Fig. 279. Various galls of *Quercus robur* and *Quercus petraea*. a) 'pineapple gall' caused by *Andricus fecundator*; b) caused by *Neuroterus numismalis*; c) 'oak apple' caused by *Biorhiza pallida*; d) caused by *Andricus lignicola*; e) 'marble gall' caused by *Andricus kollari*; f) 'spangle gall' caused by *Neuroterus quercusbaccarum*; g) caused by *Andricus curator*; h) caused by *Macrodiplosis dryobia*.

Fig. 280. *Acer* sp.
Contorted branching in a horticultural monstrosity.

A case is made (216) for the constructional organization of a flowering plant to be considered in terms of the potential, position, and time of activity of shoot apical meristems, or buds. The combined outcome of such activity will lead to the development of a branched organism. The progressive sequences of branching will be controlled internally, reflecting the form of the particular plant species, but will be flexible within limits in response to environmental fluctuations. All trees of a given species look alike because they are all conforming to a given set of branching 'rules', but each individual will have a unique array of branches reflecting its unique location and history. In order to recognize and describe the branching sequence of a plant it is useful to identify its component branching units (units of construction 282), and then the manner in which these are added to or lost from the developing structure is more readily appreciated. This section is heavily biased towards a consideration of tree construction or tree 'architecture' as it has come to be referred to latterly. Trees are large and reasonably accessible branched plants and it is the range of branching construction exhibited by tropical trees in particular that has led to a quest for knowledge of plant architecture. It is fitting to list here the publications on which this synopsis account is based: Corner 1940; Koriba 1958; Prévost (the article 286) 1967; Hallé and Oldeman (architectural models 288) 1970; Oldeman (reiteration 298) 1974; Edelin (architecutral analysis 304) 1977; Hallé *et al.* 1978; Edelin (intercalation 302, and metamorphosis 300)

1984, 1990. There is one trivial aspect of the description of branching patterns which presents constant problems: the use of imprecise terminology (see Tomlinson 1987). 'Branch' is an imprecise word. It usually implies an axis of a lesser stature to that on which it is located, and it may or may not incorporate all the lesser branches and twigs borne on it. Bough or limb usually imply something relatively big but not as big as the trunk. There is no correlation between the words available to describe the architectural and the botanical development of the structure. For example, both the trunk and a branch of a tree may be either monopodial or sympodial (250) in their make-up. If monopodial, the branch represents a shoot formed by the activity of one single apical meristem (a shoot unit 286). If sympodial, the branch represents a series of shoot units each derived from one apical meristem. This conflict between popular description and botanical detail is discussed under architectural analysis (304). In the intervening sections, loose popular terminology is employed with qualification where necessary to avoid ambiguity.

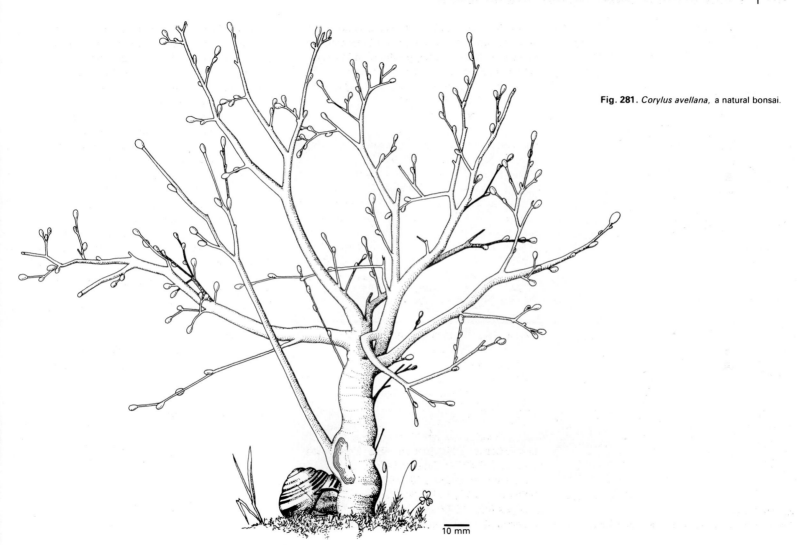

Fig. 281. *Corylus avellana*, a natural bonsai.

10 mm

Fig. 282. *Piper bicolor*
Metamers, internode plus node, of the
vertical monopodial axis; lateral branches
are sympodial and composed of series of
'articles' (286, **290b**). This species has a
winged stem (120).

A plant grows by the progressive accumulation of similar units; it is not, like most animals, a fixed shape that simply enlarges. In the study of plant developmental construction, a number of 'constructional units' real and theoretical, have been described and each has its uses depending upon the nature of morphological investigation to be undertaken. A selection of such units is listed here (282, 284); the two most appropriate to tree architecture, the article and the architectural unit, being considered in more detail elsewhere (286 and 304, respectively). A complex structure can be more readily understood if it is broken down into manageable components which can be counted and their turnover in numerical terms monitored.

A.   *Metamer* (also called a phytomer)
A metamer is a repeated constructional unit, consisting of a node plus the leaf at that node, and its subtended bud if present, plus a portion of internode (**283a–c**). A metamer may thus be deemed to include the internode proximal to it or distal to it, or a portion of each. The plant is a collection of such units, adjacent metamers possibly having similar or distinctly different morphological features (e.g. scale leaf metamer followed by foliage leaf metamer in *Philodendron* 10) (White 1984). Disruption of such a sequence will result in an abnormal plant (270; Groenendael 1985).

B.   *Phyton*
A phyton is a unit of construction representing a leaf and its node of insertion plus that portion of

stem proximal to the node into which the leaf has its vascular connections (**283d, e**). Such a segment of stem may or may not be readily identifiable by anatomical analysis. Even if it does have an identifiable anatomical reality, the concept is of dubious practical usefulness.

C.   Pipe stem model

The pipe stem model (Shinozaki *et al.* 1964) envisages a plant, such as a tree, to consist of a photosynthetic array of leaves supported and served by the trunk and branches (**283f**). Quantitatively a relationship is found between the amount (fresh or dry weight) of leaf above a given horizontal plane, and the total cross-sectional area of all stems and branches at that plane (**283f, g**). Thus the plant is seen, in theoretical terms, to consist of an assemblage of unit pipes each supporting a unit quantity of photosynthetic material. The same analysis can be applied to a stand of vegetation (**283h**). (Continued on page 284.)

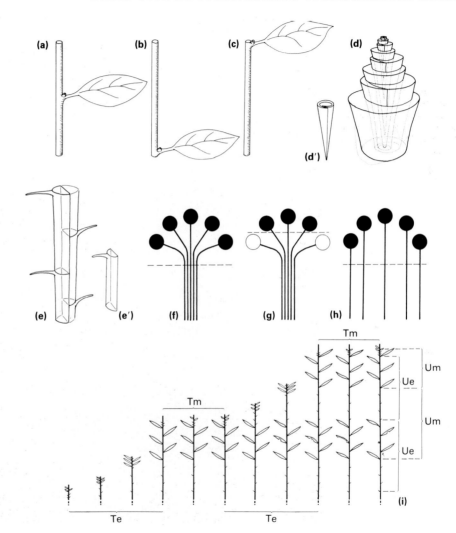

Fig. 283. a, b, c) Alternative representations of a metamer; d) a collection of phytons in a monocotyledon; d¹) single phyton from 'd'; e) a collection of phytons in a dicotyledon (distichous); e¹) single phyton from 'e'; f, g) pipe stem model of a plant; h) pipe stem model of a plant population; i) developmental sequence of *Hevea brasiliensis* (284). Te: time of extension. Tm: time of morphogenesis (leaf primordia being initiated in an apparently dormant terminal bud). Ue: unit of extension. Um: unit of morphogenesis.

D. Branch ordering

A branching system, such as a plant, can be described in terms of a hierarchy of units of successive orders. The process of applying the ordering sequence can proceed from the proximal end of the plant, e.g. the trunk of a tree out toward the ultimate distal twigs (centifugal; **285c**), or can commence at the periphery down towards the trunk (centripetal; **285e**). This latter procedure is that adopted by geographers studying river systems (e.g. Strahler 1964). When applying one of these systems the inadequacy of terminology in general use becomes apparent. Usually each branch order unit is identified subjectively as a clearly visible and unambiguous branch of whatever size (**285a**). However, in many instances a branch may be a composite structure representing a linear series of shoots each derived from a distinct apical meristem, i.e. a sympodium (250). Thus in developmental terms it is possible to describe such a single branch as a series of constructional units each of a different branch order (compare **285a** with **285b**, and **285e** with **285f**). It should be stated whether the ordering is being applied to each constructional component in botanical terms (i.e. to each module or article 286) or to the gross visible branch components (i.e. the axes 304).

E. Units of morphogenesis and units of extension

This distinction was made by Hallé and Martin (1968) in order to describe the rhythmic growth of *Hevea brasiliensis* (see 260 and **283i**).

F. Module

Plants and sedentary animals are described as being modular in their construction— built up of similar repeated units, in contrast to those animals that have a fixed but possibly enlarging shape. The module concept is applied in two distinct ways. It is used as a translation of the French *'l'article'* where it has a precise meaning in botanical morphology (286). It is also applied in a much more lax manner to mean any constructional entity that is iterated as the plant develops. In this mode, a module may represent a leaf, a metamer, a phyton, an article, a unit of branch ordering, or a ramet (170) (such as a grass tiller). It is a component of the plant that is 'born', 'dies', and can be counted. Both the leaves of a grass plant and the ramets of a grass plant are temporary structures within the lifespan of the plant (the genet) and can be monitored in terms of their population dynamics (Harper 1981, 1985).

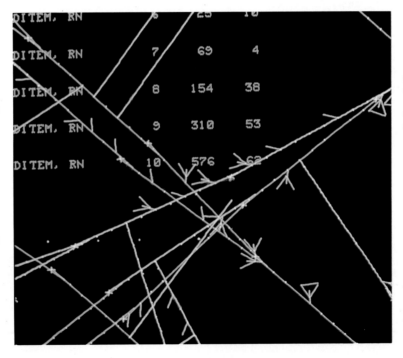

Fig. **284**. Computer graphic simulation of the growth of a modular plant (based on the growth form of rhizomes of a bamboo *Phyllostachys* sp. see Figs. **195d** and **311j**.

G. *Sympodial unit*

The distinction between a sympodial branch (sympodium) and a monopodial branch (monopodium) is made on page 250. The article (286) is a sympodial unit.

H. *Architectural unit*, page 304.

I. *Phyllomorph*

The phyllomorph (208) is a unit of construction found only in a limited number of species in the family Gesneriaceae. A similar constructional unit, the frond, is typical of the family Lemnaceae (212).

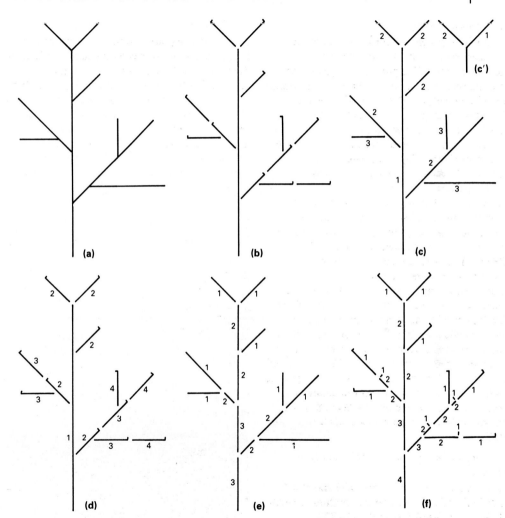

Fig. 285. Branch ordering. a) Hypothetical monopodial branching system; b) same basic system but presumed sympodial construction; c) centrifugal ordering ignoring possible sympodial construction; c¹) alternative distal arrangement; d) centrifugal ordering taking sympodial nature of the branching into consideration; e) centripetal ordering (Strahler 1964). If an order no. 1 joins an order no. 1, an order no. 2 results. If order no. 1 joins any higher order, the higher order is retained. f) As 'e', but sympodial, the evicted end of each sympodial unit having to be counted as an order.

Fig. 286. *Cyphomandra betacea*
Both the trunk and the branches are made up of sympodial units (i.e. modules or 'articles') terminating in an inflorescence. The inflorescence on the uppermost, youngest, trunk module can be seen dangling down below the uppermost tier of horizontal branches. Model of Prévost (**295i**).

The branches of a plant are either monopodial or sympodial (250). A sympodial branch is composed of a series of sympodial units which are either indeterminate (apposition growth **251e**), or are determinate in that each ceases to grow in turn, its apical meristem being converted into an inflorescence or other non-meristematic organ (substitution growth **251d**). Such sympodial units are the principle units of construction of many plants and represent one of the conspicuous features of tree architecture that is embodied in recognition of architectural models (288). For example, either the branches and/or the trunk of a tree may be built up of a developmental sequence of determinate sympodial units each derived from one apical meristem. All the units may be equivalent (**295f**) or of different non-equivalent types (**295i**) depending upon their potential. In some species the equivalence is absolute, each unit having the same number of leaves, and with buds having the same potential, time of activity, and locations. The architectural significance of the determinate sympodial unit was first expressed by Prévost (1967) in French, using the term *l'article*. *L'article*, meaning joint, is derived from the Latin *articulus*. In biology it denotes a jointed construction and was applied specifically to plant architecture by Prévost (1967) to indicate one determinate sympodial unit of a sympodium (250). '*L'article*' has been translated as 'module' in this context (Hallé *et al.* 1978), notwithstanding the existence of an old word 'caulomere' (Stone 1975) which refers specifically to a sympodial unit. As 'article' does not imply a joint in English, and 'module' has been used in a number of other connotations (284), the concept of *l'article* is represented here by the term sympodial unit. Throughout the book another term, shoot unit (or simply shoot), is employed to indicate the structure that has developed from one single apical meristem; thus a determinate sympodial unit, an indeterminate sympodial unit as in apposition growth (**251e**), and a monopodium, is each a shoot unit (250).

Fig. 287. *Piper nigrum*, flowering branch. Each inflorescence terminates a portion of stem which thus constitutes a module. I: inflorescence. Is: inflorescence scar. M: module i.e. a sympodial unit.

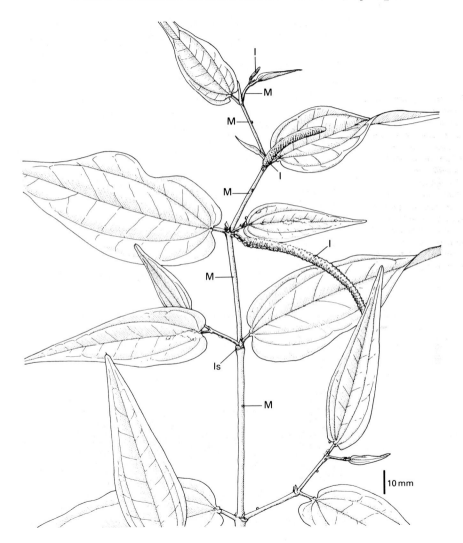

Emphasis is placed in this book on the dynamic nature of plant morphology. As the plant grows various features of its development are consistent for every individual of the species. The role of the apical meristem, or 'bud' in a looser sense (16) is highlighted (216) and details of apical meristem position, potential, and time of activity are given (218–268). Constructionally, plants consist of an accumulation of shoots, each shoot (also termed here a shoot unit 286) being produced by one apical meristem. These shoots are capable of either more or less indefinite growth (monopodial growth, the shoot being a monopodium) or are arranged in series (sympodial growth, there being a series of sympodial units, 250). Different plant species exhibit different juxtapositions of these possible components and these have been investigated and formalized in detail, particularly in the case of tropical trees (and subsequently temperate trees and other growth forms 306, 308). Observations of the sequences of events that take place during the lifespan of individuals of different tree species indicate that each species has a recognizable 'blue print' (branching sequence or model) to which the young plant conforms. The characteristic 'architectural models' for tropical trees are described by Hallé and Oldeman (1970; see also Hallé *et al.* 1978). Hallé and Oldeman initially described 24 different models, each representing a particular developmental sequence of branching. Each is labelled with the name of an appropriate scientist rather than that of a typical tree species; the latter might not be familiar world-wide. Twenty-one models were identified from pantropical rain

forests and the existence of three more predicted, examples of one of which (Stone's model **293e**) were soon found. Another model (that of McClure **295c**) was later added (Hallé *et al.* 1978). The significant point must be that this limited continuum of models has been found to be adequate to encompass the many hundreds of tree species subsequently investigated; thus the branching sequences of trees, at least, are confined within one or other of this relatively small suite of models of development (cf. intermediate forms 296). The branching exhibited by a young tree which conforms to its model may be augmented later in the life of the tree in a number of ways (reiteration 298, metamorphosis 300, intercalation 302). The 23 existing models of tree architecture are briefly itemized here. As the concept of a model incorporates a dynamic aspect, each model is presented as a simplified cartoon sequence of development (**291, 293, 295**; see also Hallé *et al.* 1978). The different models are distinguished by the presence of one or more of the following distinctive features:

- trunk monopodial (250, **289a**);
- trunk sympodial (250, **289b**);
- trunk growing continuously (260, **289c**);
- trunk growing rhythmically (260, **289d**);
- branches orthotropic (246, **289e**);
- branches sympodial and sympodial units indeterminate (i.e. branch plagiotropic by apposition 250, **289f**);
- branches plagiotropic but not by apposition (**289g, g¹**);

- flowering lateral (**289f**);
- flowering terminal (**289g**);
- the same shoot unit (286) contributing both orthotropic and plagiotropic components to the branching sequence (a 'mixed' branch; **289h, h¹**);
- the incorporation of determinate sympodial units ('article' 286) of one type only (**289i**); and
- the incorporation of determinate sympodial units of two types (**289j**).

(See pages 290, 292, 294.)

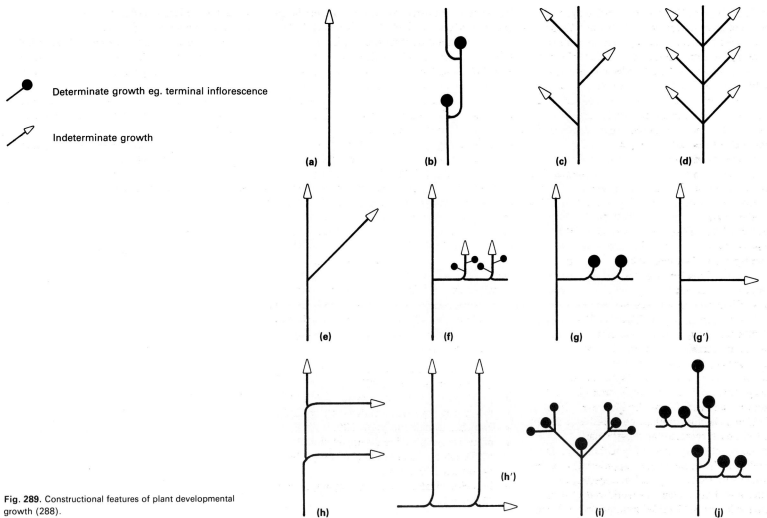

Determinate growth eg. terminal inflorescence

Indeterminate growth

(a) (b) (c) (d) (e) (f) (g) (g') (h) (h') (i) (j)

Fig. 289. Constructional features of plant developmental growth (288).

Fig. 290a. *Ficus pumila*
Model of Attims.

Fig. 290b. *Piper* sp.
Model of Petit
(photograph courtesy of
Institut Botanique,
Montpellier, France).

Model of Holttum (**291c,** Preface)
Determinate trunk; terminal inflorescence. No branches (except those within the inflorescence which are incidental to this analysis).

Model of Corner (**291d, 196**)
Monopodial (indeterminate) trunk; lateral inflorescences. No branches (except those within the inflorescences).

Model of Cook (**291e, 268b**)
Monopodial (indeterminate) trunk growing continuously. All branches temporary.

Model of Attims (**291f, 290a**)
Monopodial trunk, growing continuously. Monopodial branches orthotropic.

Model of Rauh (**291g, 250**)
Monopodial trunk, growing rhythmically. Monopodial branches orthotropic.

Model of Roux (**291h, 246**)
Monopodial trunk, growing continuously. Monopodial branches plagiotropic.

Model of Massart (**291i, 142**)
Monopodial trunk, growing rhythmically. Branches plagiotropic.

Model of Petit (**291j, 290b**)
Monopodial trunk, growing continuously. Branches composed of determinate sympodial units.

(See also pages 292, 294.)

Fig. 291. a) *Paulownia tomentosa*, model of Fagerlind but cf. Scarrone (292); b) *Phellodendron chinense*, model of Scarrone; c–j) growth models: c) Holttum, d) Corner, e) Cook, f) Attims, g) Rauh, h) Roux, i) Massart, j) Petit.

Model of Fagerlind (**293c, 291a**)
Monopodial trunk, growing rhythmically.
Branches composed of determinate sympodial
units.

Model of Aubreville (**293d, 260a, 304**)
Monopodial trunk, growing rhythmically.
Branches plagiotropic by apposition (i.e.
composed of indeterminate sympodial units).

Model of Stone (**293e, 306**)
Monopodial trunk, growing continuously.
Branches orthotropic and sympodial.

Model of Scarrone (**293f, 291b**)
Monopodial trunk, growing rhythmically.
Branches orthotropic and sympodial.

Model of Troll (with monopodial trunk;
293g, 292a)
The trunk and branches are plagiotropic except
perhaps for a short proximal portion. The
monopodial trunk is secondarily reorientated into
the vertical position by cambial activity.

Model of Troll (with sympodial trunk;
293h, 292b)
The trunk and branches are plagiotropic except
perhaps for a short proximal portion. The
proximal part of each sympodial component of
the trunk is secondarily reorientated into the
vertical position.

Model of Mangenot (**293i, 293a**)
Orthotropic sympodial trunk, the distal portion of
each sympodial unit developing sideways as a
plagiotropic branch.

Model of Champagnat (**293j, 293b**)
Orthotropic sympodial trunk, the distal portion of
each sympodial unit developing sideways and
drooping under its own weight.

(See also pages 290, 294.)

Fig. 292a. *Prunus* sp.
Model of Troll (**293g**).

Fig. 292b. *Platanus hispanica*
Model of Troll (**293h**).

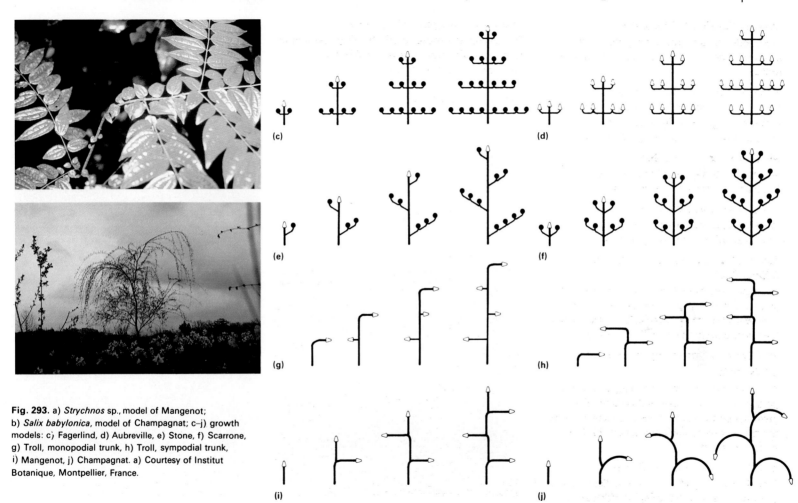

Fig. 293. a) *Strychnos* sp., model of Mangenot;
b) *Salix babylonica*, model of Champagnat; c–j) growth
models: c) Fagerlind, d) Aubreville, e) Stone, f) Scarrone,
g) Troll, monopodial trunk, h) Troll, sympodial trunk,
i) Mangenot, j) Champagnat. a) Courtesy of Institut
Botanique, Montpellier, France.

Model of McClure (**295c, 194a**)
Symodial branching sequence in which the proximal part of each determinate sympodial unit is plagiotropic and the distal part forms an orthotropic trunk. The trunk bears determinate branches.

Model of Tomlinson (**295d, 130**)
Symodial branching sequence with each orthotropic sympodial unit borne on the proximal portion of a previous unit. Sympodial units indeterminate or determinate.

Model of Chamberlain (**295e, 294a**)
Symodial trunk, no branches. Each sympodial unit bears just one similar unit at its distal end.

Model of Leeuwenberg (**295f, 294b**)
Symodial branching sequence. Each sympodial unit bears more than one similar unit at its distal end.

Model of Schoute (**295g, 295a**)
True dichotomy (258) of the apex at intervals. Flowering lateral.

Model of Koriba (**295h, 244, 266**)
Symodial trunk. Each sympodial unit of the trunk bears more than one laterally extending branch at its distal end. One of these branches is secondarily reorientated into the vertical position to become the next trunk unit (266).

Model of Prévost (**295i, 286**)
Symodial trunk. Each sympodial unit of the trunk bears more than one branch at its distal end. One of these branches is delayed in its extension and grows vertically to become the next unit of the trunk. The other branches are initially orthotropic but become plagiotropic by apposition or substitution (250).

Model of Nozeran (**295j, 295b**)
Symodial trunk. Each sympodial unit of the trunk bears more than one branch at its distal end. One of these branches is delayed in its extension and grows vertically to become the next unit of the trunk. The other branches are plagiotropic, retaining this character even if rooted as cuttings (242).

Fig. 294a. *Epiphyllum* sp.
Model of Chamberlain.

Fig. 294b. *Euphorbia punicea*
Model of Leeuwenberg.

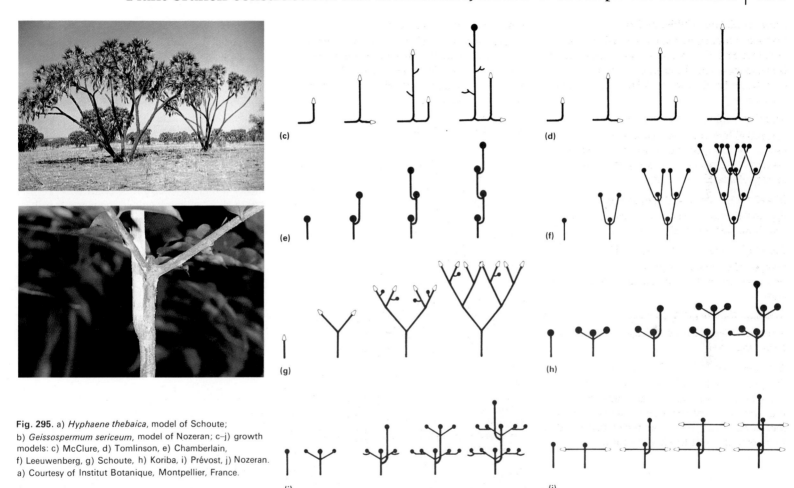

Fig. 295. a) *Hyphaene thebaica*, model of Schoute;
b) *Geissospermum sericeum*, model of Nozeran; c–j) growth
models: c) McClure, d) Tomlinson, e) Chamberlain,
f) Leeuwenberg, g) Schoute, h) Koriba, i) Prévost, j) Nozeran.
a) Courtesy of Institut Botanique, Montpellier, France.

Fig. 296. *Fremontodendron californica*
An orthotropic sympodial trunk (cf. Fig. **251a**) with orthotropic sympodial branches. This combination of branch and trunk architecture is not represented by one of the models of Hallé and Oldeman (288).

A description of tree development in terms of the series of architectural models (288–294) provides a convenient starting point for interpretation of plant form. Perhaps surprisingly most trees studied, both tropical and temperate, conform at least in their early stages of growth to one or other of the 23 branching sequences listed by Hallé *et al.* (1978). Subsequent development usually incorporates additional phenomena (298, 300, 302). It should be stressed that the precisely defined models have been considered to represent familiar and consistent points along a continuum of tree form, and it is clearly possible to construct alternative combinations of criteria leading to alternative classifications (e.g. Guédès 1982). In 1970, Hallé and Oldeman accurately predicted the likely existence of some growth sequences that they had not themselves found; their theoretical model III is now recognized as the model of Stone (**293e**), for example. Other possible models that do not feature in the existing range will doubtless be identified; *Fremontodendron californica* can develop with a sympodial trunk and with orthotropic sympodial branches (**296, 297f**) and thus represents a model or variation close to that of Prévost (**297e**). Furthermore the basic architecture of the tree may be influenced by its environment (Hallé 1978; Fisher and Hibbs 1982). An often quoted example is that of *Arbutus* species which develop according to the model of Leeuwenberg (**297g**) in the sun but develop a monopodial trunk in shade, thus qualifying for the model of Scarrone (**297h**). Such a change of model may take place inevitably as a tree ages, independently of

environmental conditions. This does not present a 'problem', it merely indicates the usefulness of a descriptive system, such as that of the concept of architectural model, which allows the unravelling of complex developmental sequences. Thus both the European sycamore (*Acer pseudoplatanus*, Aceraceae) and the tropical tree *Isertia coccinea*, Rubiaceae (Barthélémy 1986), undergo changes of branching development during their lifespan which in the context of models can be described as a switch from the model of Rauh to Scarrone to Leeuwenberg (**297a–d**). The timing only of these events will depend upon environmental conditions, principally the degree of shading. One criterion that repeatedly enters into the distinction of models is that of plagiotropy or orthotropy (246). Identification of these conditions requires more than just a recognition of horizontal or vertical growth and may be masked by shifts of branch orientation due to weight, see Fig. **247b–e**, or a gradation of morphologies (300).

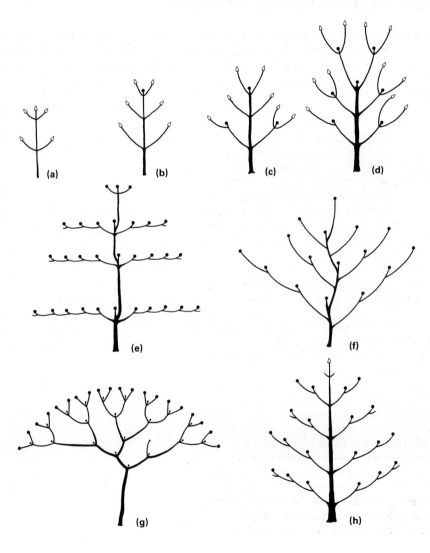

Fig. 297. a–d) A tree progressing from the model Rauh a), via the model of Scarrone b, c) (see **302**), to the model of Leeuwenberg, d); e) model of Prévost; f) *Fremontodendron* sp., both trunk and branches orthotropic and sympodial; g) *Arbutus*, open, model of Leeuwenberg; h) *Arbutus*, shade, Model of Scarrone.

The concept of an architectural model in terms of developmental morphology is outlined in section 288 with examples in sections 290–294. The model describes the generalities of branching sequence that a tree undergoes, for example, as it increases in height and matures. As the tree becomes more elaborate, it is likely that many axillary buds will not develop as part of the model, but will remain dormant, situated on the trunk or branches. One or more of these buds may subsequently be induced to develop either because the existing plant framework has been damaged, or because the plant is experiencing favourable growing conditions. In either case the branching system resulting from activation of the dormant bud will repeat in a more or less precise manner the same sequence of development as that of the parent plant developing from the seedling. The new growth forms a reiteration (Oldeman 1974) (proleptic reiteration, cf. 300) which is conforming to the same architectural model as that of its parent. Adaptive reiteration develops in response to favourable conditions (**298a, 299a**); traumatic reiteration results from injury (**298b, 299c**). Normally a reiteration is total, forming a major branch borne on the parent plant. A change in the environment of part of a tree only, such as damage or excessive light to one branch, can result in a partial reiteration (**299e**) repeating the details of the model that apply to a branch rather than to the trunk plus branch. Reiteration can also occur by a process of dedifferentiation. In this instance the apical meristem of a branch shoot changes its potential and rather than continuing to grow

plagiotropically for example, commences to grow orthotropically and exhibits the morphological characteristics for that type of shoot (**299f**). Thus it is now conforming to the total model growth rules. Such a change of potential resulting in reiteration can also occur progressively in a series of branches (metamorphosis 300).

Fig. 298a. *Alstonia scholaris*
Adaptive reiteration. Three orthotropic axes are visible, the centre one being a relatively new axis from a dormant bud. Compare with the young tree in Fig. **286** which has the same model (Prévost).

The ability to reiterate varies between tree species. In some it is practically absent even in response to damage and the individual plant always conforms to the initial expression of its model. Other species reiterate almost exclusively in response to damage only (i.e. traumatic reiteration). Perhaps a majority of species have

Fig. 298b. *Ulmus procera*
Traumatic reiterations of a damaged tree.

an ability for both traumatic and adaptive reiteration. Generally speaking the older and larger the tree becomes, the smaller in stature will be the individual reiterations. Thus in the mature crown, reiterations will occur that are greatly depauperate representations of the model. As reiteration usually occurs in response to environmental changes it is not a predictable architectural event for the plant, whereas the branching sequences conforming to the model are in a general sense predictable. Nevertheless there are indications that in some plants the reiteration process is inevitable and therefore predictable within the life span of the plant. Thus *Tabebuia* (Borchert and Tomlinson 1984) grows according to the model of Leeuwenberg (**295f**) but forms a prominent trunk by developing single reiterations one upon the other at discrete intervals of time (**299d**).

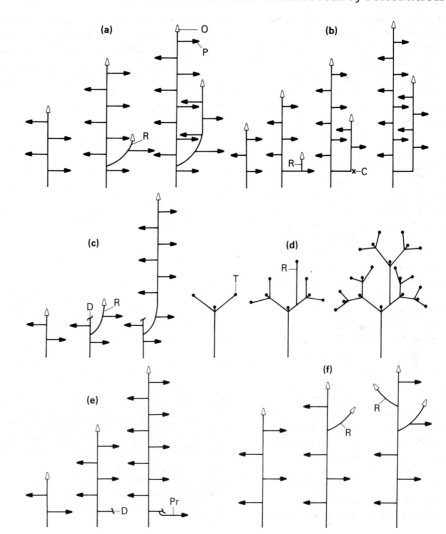

Fig. 299. Forms of reiteration (based on the model of Roux 291h). a) Adaptive reiteration from trunk bud (proleptic reiteration); b) adaptive reiteration from branch bud (lateral reiteration); c) traumatic reiteration; d) predictable reiteration; e) partial reiteration; f) dedifferentiation. C: cladoptosis (268). D: damage. O: orthotropic shoot. P: plagiotropic shoot. Pr: partial reiteration. R: reiteration. T: terminal inflorescence.

The progressive construction of a tree has been interpreted in terms of three interlocking concepts, that of the architectural model (290–294), that of the repetition of the model (reiteration 298) and that of a change in the potential of a branch from a generally plagiotropic (246) disposition to a generally orthotropic (246) disposition. This latter change is termed metamorphosis (Hallé and Ng 1981; Edelin 1984, 1990). The process of metamorphosis has been recorded for a number of different tree species and is doubtless a constant feature of a great many more yet to be investigated. The change in branch apical meristem potential can be abrupt, that is to say one branch may be plagiotropic but the next highest branch will be orthotropic (**301a**). If the sequence of metamorphosis is gradual, it can be expressed in two ways. A transition zone can exist (**301b**) in which each branch is plagiotropic proximally but has an orthotropic distal end. The higher in the transition zone the greater will be the proportion of orthotropic axes. Alternatively, successive branches exhibit a gradual loss of plagiotropic morphological characters and a concurrent gain in orthotropic characters (such as loss of opposite phyllotaxis and gain of spiral phyllotaxis 224, **301c**). Thus the expected potential for buds along the trunk changes and is under some sort of internal control. By whatever means metamorphosis proceeds, the orthotropic components may each exhibit the total potential of the model and thus represent reiterations (298). This process is often accompanied by an increase in the intensity of branching such that

reiteration complexes develop. These complexes, derived by the process of metamorphosis, are termed sylleptic reiteration complexes (**301d**) to distinguish them from reiteration derived from dormant buds (proleptic reiteration 298). The process of metamorphosis will affect the whole tree to differing extents. All second order axes (see architectural analysis 304), i.e. branches attached directly to the trunk, may be involved, except perhaps the lowest and oldest (e.g. **301a**), and will then be equivalent to the A1 type axis in an architectural plan (304); or the occurrence of metamorphosis may be more diffuse and less easy to identify (**301e**). It may affect only the second order axes, higher orders remaining plagiotropic (**301f**). Alternatively it may affect each order in

turn, the third order axes undergoing progressive metamorphosis in one of the manners described above only once all the second order axes have completed the change (see intercalation 302). There can be a considerable pause when normal growth resumes between each metamorphosis event affecting successive orders of axes. In summary, trees exist in which metamorphosis is apparently absent; at the other extreme the phenomenon is extensive, accompanied by a proliferation of branching orders, each undergoing a metamorphosis sequence and giving rise to sylleptic reiteration complexes. Yet other trees undergo metamorphosis but do not form these extensive systems.

Fig. 300. *Maesopsis eminii*
Metamorphosis in the model of Roux (**291h**). Courtesy of Institut Botanique, Montpellier. France.

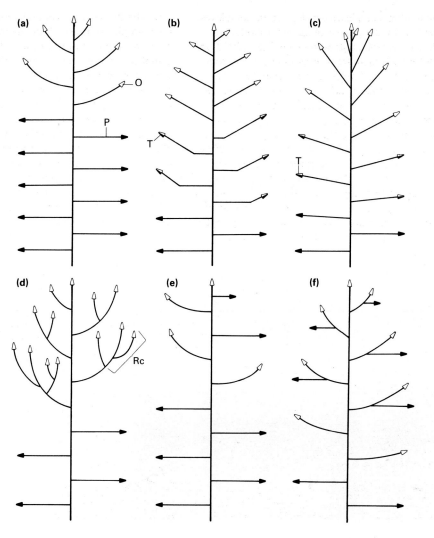

Fig. 301. Forms of metamorphosis. O: orthotropic shoot. P: plagiotropic branch. Rc: reiteration complex. T: transition branch.

The process of intercalation (Edelin 1984) takes place during the increase in complexity of a developing tree as the light-intercepting periphery of the tree is situated further and further from the trunk. The tree can be visualized as being composed of three zones (**303a–c**): the supportive trunk (1), the peripheral zone of the canopy (3), usually intercepting most light and likely to be reproductive, and an intermediate zone (2) which provides a support structure bridging the gap between the periphery and the trunk in the larger tree. Zones (1) and (3) always exist, zone (2) is interspersed or intercalated between these two at a later developmental stage. In Fig. **303d–g** representing the model of Roux (**291h**), the young tree, conforming strictly to the model, has a monopodial orthotropic trunk and monopodial plagiotropic branches. Later in its development, newly produced branches are oblique (representing the process of metamorphosis 300) and are intercalated between zone (1) and zone (3). Thus the branches of the crown edge are still plagiotropic. This process will continue further up the trunk, and more and more orders of branches will be intercalated between trunk and periphery (**303h**). This sequence is by no means indefinite; depending on the species, increase in branching will soon be due to proleptic reiterations rather than intercalation (**303i**).

Fig. 302. ***Acer pseudoplatanus*** Intercalation. (Note the fork at the top of the trunk, see **297a–d**.)

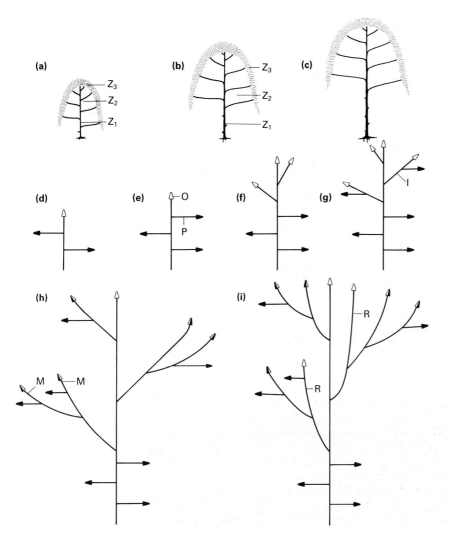

Fig. 303. Process of intercalation. I: intercalation. M: metamorphosis. O: orthotropic shoot. P: plagiotropic shoot. R: reiteration. Z_1: zone 1, supporting trunk. Z_2: zone 2, intermediate zone. Z_3: zone 3, peripheral zone.

Fig. 304. *Sterculia* sp.
Model of Aubreville (**293d**). Monopodial trunk. Branches
sympodial by apposition (**250**). See small slender
orthotropic distal ends of sympodial branch units on upper
surface of massive branch.

It is not necessarily possible or indeed particularly useful to describe the details of branching of one individual tree in totality. Rather, progress has been directed towards describing the general rules of branch growth of a tree species, the general but predictable sequence of development to be expected from any representative individual, bearing in mind that each individual will be unique in its own branching detail due to its unique local environment. The branching development of a tree (or other plant form 306, 308) exhibits a number of morphological features: the architectural model (288), reiteration (298), metamorphosis (300), and intercalation (302). What is needed to bring these together and to encapsulate, in essence, the sequence of developmental events that proceed from germinating seed to senile tree is a synopsis of these events, or an architectural analysis (Edelin 1977, 1984, Barthélémy *et al.* 1989a) (see also age states 314). The fundamental point to note is that for any one species there will be a finite number of branch categories collectively constituting an *architectural unit*. In order to avoid the possible ambiguities of 'branch' (280), the categories are referred to as axis type 1 (always the trunk or major trunk component for a tree), axis type 2, axis type 3, etc. These are not necessarily the same as branch order 1, 2, and 3 (284) as the outermost axis type— probably bearing the leaves—will be borne directly on the trunk (axis type 1) of a very young tree but borne perhaps on axis type 3 or 4 in an older tree (**303d–i**). Each species will have a different set (architectural unit) of axes types.

Thus in the hypothetical example shown (**305a**), the architectural unit consists of three categories of branch each defined in terms of dynamic morphology, i.e. potential, position, and time of activity and which is summarized in an architectural plan (**305g**). These are the constructional building components that will in unison constitute the framework of the tree. They will be juxtaposed in accordance with the model (288) of the tree, modified in many cases by the process of metamorphosis (300) whereby for example a type 2 axis will take on the characteristics of a type 1 axis. Depending upon the species, the sequence of branching they represent will be replicated within the overall tree framework by the process of reiteration (298). By means of an architectural unit, sequential diagrams of the branching development (**305c–f**), and observations concerning the occurrence of reiteration and metamorphosis, a general synopsis of the plant's form is possible. To arrive at this synopsis, it will be necessary to observe plants of different ages and to test it out on additional plants of a range of ages and locations in order to confirm its validity (Edelin 1990).

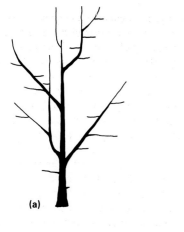

(a)

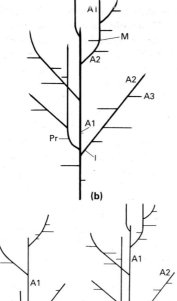

(b)

(c) A1

(d) A2 A1 A2 A3

(e) A1 A1 S

(f) A1 A2 A3

(g)

Axis 1	**Axis 2**	**Axis 3**
Monopodial	Monopodial	Monopodial
Continuous growth	Continuous growth	Continuous growth
Orthotropic	Orthotropic	Plagiotropic
Spiral phyllotaxis	Spiral phyllotaxis	Distichous phyllotaxis
Non-sexual	Non-sexual	Sexual
Non-shedding	Non-shedding	Shedding
Large scale leaves	Small scale leaves	Foliage leaves
Indeterminate	Determinate in the long term	Determinate

Fig. 305. Architectural analysis. a) Hypothetical tree; b) simplified diagram of 'a'; c–f) developmental sequence of 'b'; g) hypothetical architectural plan for sequence c–f. A1: axis type 1. A2: axis type 2. A3: axis type 3. I: intercalation of A2 between A3 and A1. M: metamorphosis of A2 into an A1. Pr: proleptic reiteration. S: self pruning (cladoptosis 268) of A3.

Fig. 306. *Rhipsalis bambusoides*
A species of cactus (202) having a monopodial main axis with modular (284) side branches conforming to the architectural tree model of Stone (**293e**).

A system for analysing tree form as expressed in terms of the development of branching pattern, is contained in sections 288–304. This approach is also applicable to the study of branching in herbaceous plants (Jeannoda-Robinson 1977). Indeed, the models of Tomlinson and McClure (**295d, c**) are more applicable to herbaceous than arborescent plant types. Many herbaceous plants have a branching system that can be directly compared to one of the architectural models of Hallé and Oldeman (1970). Many tussock forming plants, for example, represent the model of Leeuwenberg (**295f**). In herbaceous plants the orthotropic components of branching are lacking to an extent and there is an emphasis on plagiotropic growth. In stoloniferous plants there may be an extensive plagiotropic monopodial system bearing orthotropic axes at intervals or the whole system may be sympodial with the proximal end of a sympodial unit being plagiotropic and its distal end orthotropic (**289h¹**). These various combinations may or may not lend themselves to irrefutable recognition as one model or another. The mode of development of a trunk of a tree is obviously an important factor in determining its model but this is a feature that is missing from a herbaceous plant. Herbaceous plants definitely exhibit the phenomenon of reiteration (298). Whether those of metamorphosis (300) and intercalation (302) are also applicable awaits consideration. Metamorphosis is a process that leads to the establishment, in a tree, of a more or less determinate crown. This may apply to shrubby herbs but probably not to indeterminate clonal plants which spread indefinitely. Jeannoda-Robinson (1977) categorizes herbs as follows:

(1) conformation to a model (**307a, b, 306**);

(2) conformation to a model in a prostrate form (**307d–f**);

(3) conformation to part of a model branching sequence (**307c**); and

(4) exhibiting some novel branching sequence.

It is likely that a thorough analysis of herb architecture, in the manner of that attempted for trees (304), could lead to an equivalent system for deciphering their developmental branching and that many aspects are possibly common to both. Herbaceous plants live at or near ground level and their development is directed towards horizontal rather than vertical growth; this aspect is reflected in their branching behaviour (310).

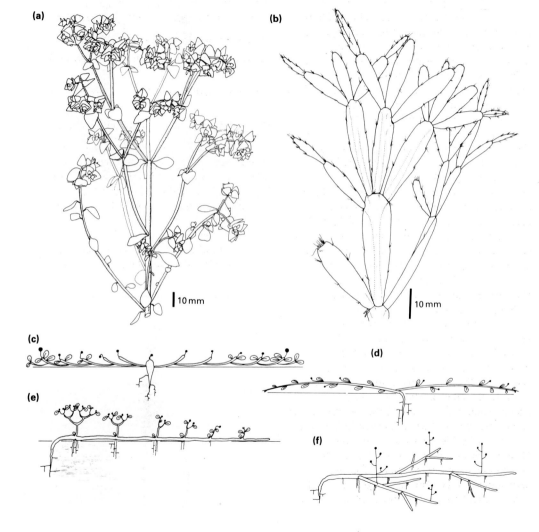

(a)

(b)

(c)

(d)

(e)

(f)

10 mm

10 mm

Fig. 307. a) *Euphorbia peplus* (model of Rauh **291g**); b) *Rhipsalidopsis rosea* (model of Leeuwenberg **295f**); c) partial representation of model of Prévost (**295i**); d) prostration of model of Troll (**293g**); e) prostration of model of Stone (**293e**); f) prostration of model of Attims (**291f**). c–f after Jeannoda-Robinson (1977).

The identification of models of architectural development that are applicable to the branching patterns of trees (288) and herbs (306) is also possible for other growth forms. Thus woody climbing plants (lianes), having representatives in a wide range of unrelated families, nevertheless exhibit a limited range of patterns of branching. To some extent these patterns are recognizable as those that are found in tree architecture, but with distinct morphological features that reflect the climbing habit, such as the production of terminal tendrils in the model of Leeuwenberg (**309a**) in positions where inflorescences would occur in a free-standing tree. Other recurring branching themes appear to be confined to lianes and do not occur in trees. A consistent feature of liane architecture is that there is a distinct juvenile and adult form (cf. establishment 168, age states 314). Juvenile forms may be free-standing and slow growing in contrast to the adult form. Conversely the juvenile form (or a reiterative shoot 298) may be represented by stoloniferous or rhizomatous shoots, the distal ends of which form the adult climbing form given a suitable support. Cremers (1973, 1974; also reported in Hallé *et al.* 1978; Cabellé 1986) records lianes from the Ivory Coast, Africa, conforming to thirteen of the tree models described by Hallé and Oldeman (1970), viz.: Corner (**291d**), Tomlinson (**295d**), Chamberlain (**295e**), Leeuwenberg (**295f**), Schoute (**295g**), Petit (**291j**, **309b**), Nozeran (**295j**), Massart (**291i**), Roux (**291h**, **309c**), Cook (**291e**), Champagnat (**293j**), Mangenot (**293i**), and Troll (**293g**, **h**), and representing twenty-five species in fifteen families. A further eleven species exhibit architectures not found in trees and are of three basic types:

(1) juvenile form orthotropic (246); adult climbing form monopodial with lateral inflorescences (**309d**);
(2) juvenile form orthotropic; adult climbing form sympodial (**309e**); and
(3) juvenile form plagiotropic (246) then climbing by means of adventitious roots (**309f**).

Fig. 308. *Bauhinia* sp. Proximal end of liane (cf. **121c**).

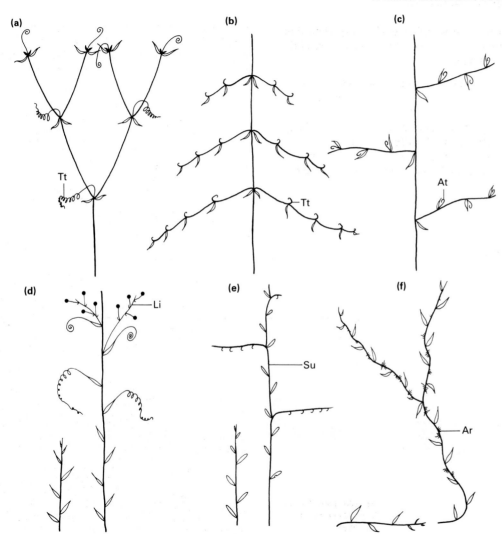

(a)

(b)

(c)

Tt

Tt

At

(d)

Li

(e)

Su

(f)

Ar

Fig. 309. a) Model of Leeuwenberg (**295f**); b) model of Petit (**291j**), c) model of Roux (**291h**); d) monopodial with lateral inflorescences; e) sympodial; f) climbing by adventitious roots. Ar: adventitious root. At: axillary tendril. Li: lateral inflorescence. Su: sympodial unit. Tt: terminal tendril.

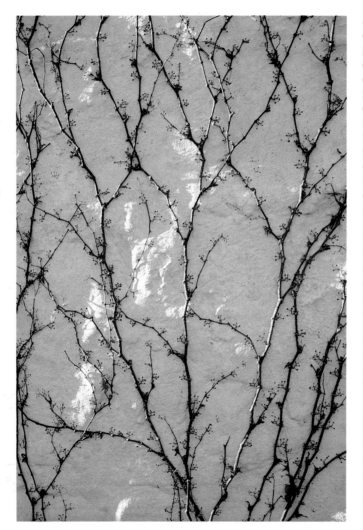

Fig. 310. *Parthenocissus tricuspidata*
Growth of this plant on a flat rock face (or wall) reveals a consistency of branching. Picking up the sequence at any point it will be found that any two tendrils on the same side of the stem will invariably have a bud or branch growing out on the same side between them (see 122 and **229b**).

A certain emphasis has been placed in this book on the dynamic nature of a flowering plant. A plant is not a static object but is constantly changing its shape by the addition of new components and by the loss of others. Plant morphology, as a discipline, has tended to concentrate on the description of plant organs rather than on consideration of the elaborating whole. However, in order to appreciate plant morphology fully the dynamic aspects must be considered from two standpoints. Firstly, plant form can be interpreted more readily if the developmental sequences of events are recognized rather than too much emphasis being placed on the mere description of organs in isolation. This is equally true of development in detail (18, 94, 112) and of the development of whole organisms (280, 304). Secondly, a plant is dynamic in the sense that it grows, it extends into the surrounding environment. In this context the comparison has been made from time to time that form, or more precisely changes in form due to growth, is for a plant the equivalent of behaviour in an animal. Thus, Arber (1950) suggests "Among plants, form may be held to include something corresponding to behaviour in the zoological field . . . for most though not for all plants the only available forms of *action* are either growth, or discarding of parts, both of which involve a change in the size and form of the organism." As a generalization, expressly excluding colonial sedentary animals, an animal does not change its shape but moves about foraging for food in accordance with a pattern of behaviour. Correspondingly, a plant departs from its germination site by growth which usually involves the process of branching, the branches maintaining and extending a photosynthetic display of leaves. The control of branching (288–294) producing this display can then be compared with the foraging behaviour of an animal. Clear examples of movement due to growth are provided by rhizomatous plants (130) in which extending growth at the distal ends is matched by death and disintegration at the proximal ends resulting in mobile organisms. In a tree, long shoots (**254**) have been described as exploring the environment, while the short shoots they bear are located along their length, exploiting the environment already visited. If plant form, expressed in terms of patterns of branching, is considered to be equivalent to the foraging behaviour of animals (Bell 1984; Sutherland and Stillman 1988), then it is tempting to look for efficiency in the pattern of branching (312). Certainly it must be significant that a relatively conservative range of branching patterns is apparent in the plant kingdom (288) and that similar patterns recur in unrelated plants (**311**). A full discussion of growth movements is given in Hart, 1989, and a discussion of plant behaviour in Silvertown and Gordon, 1989. Adjustment of branching pattern following the sensing of distant neighbours is reported by Novoplansky *et al.* 1990.

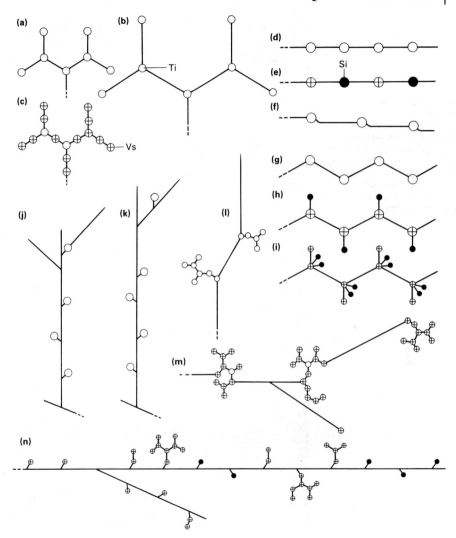

Fig. 311. Diagrammatic plan view of rhizome branching systems of bamboos (a,, b, j, k, l, m) and gingers (a, b, c, d, e, f, g, h, i, l, m). Si: solitary inflorescence or flower. Ti: terminal inflorescence. Vs: vegetative shoot. Redrawn in part from Bell and Tomlinson (1980).

Fig. 312a. *Philodendron* sp.
Shadows on the tree trunk indicate the effective functioning of pulvini (46) in leaf orientation.

Fig. 312b. *Qualea* sp.
Crown shyness. Individual branch clusters cease growing before they become entangled with their neighbours.

It is always tempting to ascribe a function to each and every morphological feature of a plant (Givnish 1983). There is no denying the performance and thus function of a tendril on a climbing plant. However, there are many features of which the functions are not at all clear, or to which suggested functions are not testable in an irrefutable manner. For example, leaves of the family Leguminosae have stipules (52). These may be represented by spines (**57a, f**) and then are quite reasonably considered to have a protective function, whereas in other species the stipules are ephemeral (**80b**) and represent merely an inevitable consequence of the mode of development of a leguminous leaf. These general observations apply to a consideration of branching patterns exhibited by plants. In many plants the branching pattern is visually precise, often geometrically remarkably accurate, and predictable (**229**). In other plants pattern is not detectable (Maddox *et al.* 1989), even by means of statistical analysis (Schellner *et al.* 1982, Cain 1990). However, when precise patterns are found it is difficult to avoid formulating the hypothesis that they represent an efficient system, efficient in terms of a balance between economic production of mechanical supporting tissue and an adequately functioning display of perhaps leaves or flowers or roots (310). The functioning of some branching systems found in nature is tantalisingly obvious. The ciliated feeding grooves on the surface of a crinoid form a pattern conveying food particles to the central mouth that is precisely the same as the optimum layout for a plantation roadway system designed

to efficiently convey bunches of bananas from surrounding fields to a central factory (Cowen 1981). Precision of branching in plants may be represented by total conservatism in the number of parts such as the number of pairs of leaves on a determinate sympodial unit (286), or represented by consistency of orientation: the rhizomes of many plants (**269d**) have zigzag sequences of sympodial units in which a left-hand unit inevitably bears a unit to the right and vice versa.

Branch lengths and angles can also be remarkably precise and are most readily seen in many inflorescences (**142**). Having identified a pattern of branching, speculation will lead to a presumed function which may or may not be a reality. Again it should be repeated that without controlled experimentation it is not possible to state that a particular display of leaves, for example, located by a branching pattern, is functioning efficiently in the interception of light rather than in the cooling of surfaces, or the resistance to wind, or shedding of snow, or shading of competitors, or is associated with display of flowers. Nevertheless investigations of branching pattern efficiency are worthwhile so long as the snags are appreciated. If the hypothesis is that a particular pattern displayed by a plant is 'efficient' then this efficiency should ideally be compared with similar but inefficient examples which will not, presumably, be extant! One solution is to utilize computer graphic simulation (Fisher and Honda 1979*a, b*, Prusinkiewicz *et al.* 1988). The actual pattern can be created, ideally with accuracy of botanical development (**216**) (Reffye *et al.* 1988, 1989), and tested for efficiency in whatever category is considered appropriate. A continuum of alternative patterns based on the actual plant can be treated similarly, the hypothesis being that they will prove to be less efficient. A number of such investigations is recorded in Bell (1986); an example is illustrated here from the investigations of Borchert and Tomlinson (1984) on the display of whorls of leaves in the tree *Tabebuia rosea* (see also reiteration **299d**).

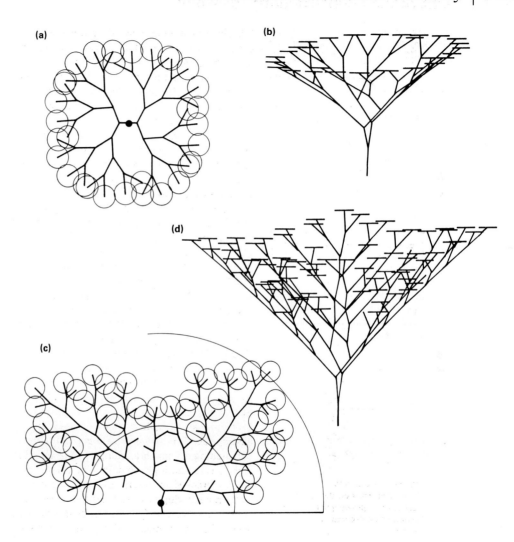

Fig. **313**. Computer simulation of symmetric and asymmetric branching in *Tabebuia* (model of Leeuwenberg **295f**). a) Plan view of leaf display, symmetrical branching (circles represent whorls of leaves); b) side view of 'a'; c) plan view of leaf display, asymmetrical branching; d) side view of 'c'. In the young plant a) leaves are restricted to the crown periphery. The onset of asymmetric branching in older plants c) results in a more even display of leaves. Subsequently a predictable reiteration will take place (**299d**), forming an additional tier of branches. The plant thus exhibits a sequence of branching strategies (**314**). Redrawn from Borchert and Tomlinson (1984). See also Borchert and Honda (1984).

An understanding of the various parts of flowering plants (1–212) allows the composite form of the plant to be considered in the broader context of its place in the environment over time. The second half of the book takes one central theme, the recognition that the plant is a dynamic organism forever augmenting and modifying its shape (growth habit or growth form). The detailed activity (position, potential, and timing) of 'buds' (16) is chosen as the key component of this morphological sequence. However, the concept of the form of the plant can be approached in other complementary ways. The plant can be considered in terms of its morphological emphasis in construction between 'root' and 'shoot' (Groff and Kaplan 1988). The plant can be considered in terms of its 'life form' (Raunkiaer 1934). The main categories of Raunkiaer's life forms are based on the stature of the plant and particularly the extent to which it is modified by seasonal conditions (especially cold or drought). A much elaborated key is provided by Ellenberg and Mueller-Dombois (1967) in which non-seasonal plants are accommodated and categories are subdivided at length. Thus geophytes (**315d**) are subdivided into root budding, bulbous, rhizomatous, or aquatic geophytes. Thallo-chamaephytes (i.e. non-vascular plants living at or near ground surface) are initially divided into hummock-forming mosses, cushion-forming mosses, and cushion-forming lichens. 'Tree' (phanerophytes **315a**) features include height, crown shape, leaf size and shape, rooting features above ground, and bark. In this manner any plant or plant community can be categorized by a combination of morphological attributes.

Raunkiaer life form analysis does not take into account the change in morphology of a plant over time, whereas this aspect is the central theme of tree form analysis (288–302) and also the key feature of the recognition of life styles (Gatsuk *et al.* 1980) represented by architectural changes within the individual plant: plants that spread widely and break into parts, plants that break up towards the last stages of their lifespan, and plants that remain intact until they die. In addition, the architecture of any plant identifies it as being representative of one stage (or 'state') of a continuum from germination to death. These states are seed, seedling, juvenile, immature, virginile, reproductive (young, mature, old), subsenile, and senile. Each state is recognized by details of morphology (and reproductive potential). In some plants the morphological

Fig. 314. Contrasting growth forms sharing the same habitat, Mexico.

changes from one stage to the next are well marked and quite abrupt and are most pronounced during the early stages of development, from seedling to juvenile, when the plant is becoming established (168) and may well exhibit features that will not be seen again during its life time, such as monopodial rather than sympodial growth (250). The age state of a plant reflects its state of development and is totally independent of calendar age; a tree seedling in an open habitat may be 1 year old and rapidly approaching the juvenile state, whilst another individual of the same species growing in a closed habitat may be many years old, and held at the seedling state until light conditions improve. Differences in behaviour of the plant (310) as represented by its dynamic morphology occur at different stages in its life and environmental circumstances; thus the immature *Philodendron* described in Fig. 11 changes its morphology once it begins to climb (**66a**), temporarily ceasing to produce foliage leaves but having much extended hypopodia (12, **315h–l**). Some examples for trees are given by Barthélémy *et al.* 1989. The concept of age state has also been referred to as biological age, ontogenetic age, and physiologic age.

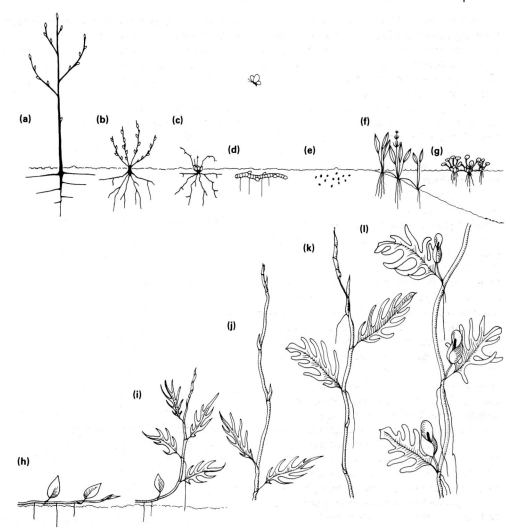

Fig. 315. a–g) The simplest subdivisions of Raunkiaer's Life Forms: a) phanerophyte, b) chamaephyte, c) hemicryptophyte, d) geophyte, e) therophyte, f) helophyte, g) hydrophyte. h–l) Probable age states of *Philodendron pedatum* (cf. **11**, **66a**): h) seedling (horizontal with simple leaves), i) juvenile (lobed leaves, short internodes), j) immature (climbing with aborted foliage leaves), k) virginile (climbing with elaborate leaves), l) adult (flowering).

Arber, A. (1950). *The natural philosophy of plant form*. Cambridge University Press, Cambridge.

Arber, A. (1954). *The mind and the eye. A study of the biologist's standpoint*. Cambridge University Press, Cambridge.

Barlow, P. W. (1986). Adventitious roots of whole plants: their forms, functions, and evolution. In *New root formation in plants and cuttings,* (ed. M. B. Jackson), pp. 67–110. Nijhoff-W-Junk, The Hague.

Barthélémy, D. (1986). Establishment of modular growth in a tropical tree: *Isertia coccinea* Vahl. (Rubiaceae). *Philosphical Transactions of the Royal Society, London*, B313, 89–94.

Barthélémy, D. (1987). Une mode de développement remarquable chez une orchidée tropicale: *Gongora quinquenervis* Ruizet Pavon. *Comptes Rendus de l'Académie des Sciences*, 304, Series III, No. 10, 279–84.

Barthélémy, D., Edelin, C. and Hallé, F. (1989a). Architectural concepts for tropical trees. In *Tropical forests. Botanical dynamics, speciation and diversity*, (ed. L. B. Holm-Nielsen, I. C. Nielsen, and M. Balslev), pp. 89–100. Academic Press, London.

Barthélémy, D., Edelin, C. and Hallé, F. (1989b). Some architectural aspects of tree ageing. In *Forest Tree Physiology* (ed. E. Dreyer *et al.*), *Annales Scientifiques Forestieres*, 46, supplement, 194s–198s. Elsevier/INRA.

Bell, A. D. (1974). Rhizome organization in relation to vegetative spread in *Medeola virginiana*. *Journal of the Arnold Arboretum*, 55, 458–68.

Bell, A. D. (1984). Dynamic morphology: a contribution to plant population ecology. In *Perspectives on plant population ecology*, (ed. R. Dirzo and J. Sarukhán), pp. 48–65. Sinauer, Sunderland, Mass.

Bell, A. D. (1986). The simulation of branching patterns in modular organisms. *Philosophical Transactions of the Royal Society, London*, B313, 143–59.

Bell, A. D. and Tomlinson, P. B. (1980). Adaptive architecture in rhizomatous plants. *Botanical Journal of the Linnean Society*, 80, 125–60.

Bell, P. R. (1985). Introduction in *The mind and the eye. A study of the biologist's standpoint* by Agnes Arber (1954), reissued 1985 with an Introduction. Cambridge Classic Series, University of Cambridge.

Bierhorst, D. W. (1971). *Morphology of vascular plants*. Macmillan, NY.

Borchert, R. and Honda, H. (1984). Control of development in the bifurcating branch system of *Tabebuia rosea*: a computer simulation. *Botanical Gazette*, 145, 184–95.

Borchert, R. and Tomlinson, P. B. (1984). Architecture and crown geometry in *Tabebuia rosea* (Bignoniaceae). *American Journal of Botany*, 71, 958–69.

Caballé, G. (1986). Sur la biologie des lianes ligneuses en forêt Gabonaise. Thèse Docteur d'Etat. Université des Sciences et Techniques du Languedoc, Montpellier, France.

Cain, M. L. (1990). Models of clonal growth in *Solidago altissima*. *Journal of Ecology*, 78, 27–46.

Cannon, W. A. (1949). A tentative classification of root systems. *Ecology*, 30, 542–8.

Charlton, W. A. (1968). Studies in the Alismataceae. I. Developmental morphology of *Echinodorus tenellus*. *Canadian Journal of Botany*, 46, 1345–60.

Cook, R. E. (1988). Growth of *Medeola virginiana* clones. I. Field observations. *American Journal of Botany*, 75, 725–31.

Cooney-Sovetts, C. and Sattler, R. (1986). Phylloclade development in the Asparagaceae: an example of homoeosis. *Botanical Journal of the Linnean Society*, 94, 327–71.

Corner, E. J. H. (1940). *Wayside trees of Malaya in two volumes*. The Government Printing Office, Singapore.

Corner, E. J. H. (1946). Suggestions for botanical progress. *New Phytologist*, 45, 185–92.

Corner, E. J. H. (1964), *The life of plants*. University of Chicago Press, Chicago.

Cowen, R. (1981). Crinoid arms and banana plantations: an economic harvesting analogy. *Paleobiology*, 7, 332–43.

Cremers, G. (1973). Architecture de quelques lianes d'Afrique Tropicale, 1. *Candollea*, 28, 249–80.

Cremers, G. (1974). Architecture de quelques lianes d'Afrique Tropicale, 2. *Candollea*, 29, 57–110.

Cullen, J. (1978). A preliminary survey of ptyxis (vernation) in the the Angiosperms. *Notes from the Royal Botanic Garden, Edinburgh*, 37, 161–214.

Cusset, G. and Cusset, C. (1988). Etude sur les Podostemales, 10. Structures florales et végétatives des Tristichaceae. *Bulletin du Muséum National d'Histoire Naturelle*, Section B, Adansonia No. 2, 4th Serie, 10, 179–218.

Cutter, E. G. (1971). *Plant anatomy: experiment and interpretation, Part 2, Organs*. Edward Arnold, London.

Darwin, C. (1875). *Insectivorous plants*. John Murray, London.

Darwin, C. (1884). *The different forms of flowers on plants of the same species*. John Murray, London.

Daubs, E. H. (1965). *A monograph of Lemnaceae*, Illinois Biological Monographs, 34, University of Illinois Press, Urbana.

Davey, A. J. (1946). On the seedling of *Oxalis hirta* L. Annals of Botany, 39, 237–56.

Davis, P. H. and Cullen, J. (1979). *The identification of flowering plant families*. Cambridge University Press, Cambridge.

Dengler, N. G., Dengler, R. E., and Kaplan, D. R. (1982). The mechanism of plication inception in Palm leaves: histogenetic observations on the pinnate leaf of *Chrysalidocarpus lutescens*. *Canadian Journal of Botany*, 60, 2976–98.

Dickinson, T. A. (1978). Epiphylly in angiosperms. *The Botanical Review*, 44, 181–232.

Dressler, R. L. (1981). *The orchids*. Harvard University Press, Cambridge, Mass.

Eames, A. J. (1961). *Morphology of the angiosperms*. McGraw-Hill, NY.

Edelin, C. (1977). Images de l'architecture des conifères. Thèse, Docteur de Specialité de Sciences Biologiques. Université des Sciences et Techniques du Languedoc, Montpellier, France.

Edelin, C. (1984). L'architecture monopodiale: l'example de quelques arbres d'Asie tropicale. Thèse, Docteur d'Etat. Université des Sciences et Techniques du Languedoc, Montpellier, France.

Edelin, C. (1990). The Monopodial Architecture: The Case of Some Tree Species from Tropical Asia. FRIM Research Pamphlet No. 105, ISSN: 0126-8198, Forest Research Institute Malaysia Publication.

Eiten, L. T. (1976). Inflorescence units in the Cyperaceae. *Annals of the Missouri Botanical Gardens*, 63, 81–112.

Ellenberg, H. and Mueller-Dombois, D. (1967). A key to Raunkiaer plant life forms with revised subdivisions. *Ber. Geobot. Inst. Eidg. Tech. Hochsch. Stift. Rubel*, 37, 56–73.

Esau, K. (1953). *Plant anatomy*. Wiley, New York, London.

Faegri, K. and Pijl, L. van der (1979). *The principles of pollination ecology*, (3rd edn). Pergamon Press, Oxford.

Fisher, J. B. and Hibbs, D. E. (1982). Plasticity of tree architecture: specific and ecological variations found in Aubreville's model. *American Journal of Botany*, 69, 690–702.

Fisher, J. B. and Honda, H. (1979a). Branch geometry and effective leaf area: a study of *Terminalia*-branching pattern. 1. Theoretical trees. *American Journal of Botany*, 66, 633–44.

Fisher, J. B. and Honda, H. (1979b). Branch geometry and effective leaf area: a study of *Terminalia*-branching pattern. 2. Survey of real trees. *American Journal of Botany*, 66, 645–55.

Fisher, J. B. and Stevenson, J. W. (1981). Occurrence of reaction wood in branches of dicotyledons and its role in tree architecture. *Botanical Gazette*, 142, 82–95.

Fitter, A. H. (1982). Morphometric analysis of root systems: application of the technique and influence of soil fertility on root development in two herbaceous species. *Plant, Cell and Environment*, 5, 313–22.

Foster, A. S. and Gifford, E. M. (1959). *Comparative morphology of vascular plants*. W. H. Freeman, San Francisco.

Gatsuk, L. E., Smirnova, O. V., Vorontzova, L. I., Zaugolnova, L. B., and Zhukova, L. A. (1980). Age states of plants of various growth forms: a review. *Journal of Ecology*, 68, 675–96.

Gerard, J. (1633). *The herball or generall historie of plants*. London.

Gerrath, J. M. and Posluszny, U. (1988). Morphological and anatomical development in the Vitaceae. 1. Vegetative development in *Vitis riparia*. *Canadian Journal of Botany*, 66, 209–24.

Gerrath, J. M. and Posluszny, U. (1989). Morphological and Anatomical development in the Vitaceae. III. Vegetative development in *Parthenocissus inserta*.

Gifford, E. M. and Foster, A. S. (1989). *Morphology and Evolution of Vascular Plants*. W. H. Freeman, Oxford.

Givnish, T. J. (1983). Introduction (pp. 1–9) to: On the economy of plant form and function. In *Proceedings of the Sixth Maria Moors Cabot Symposium*, (ed. T. J. Givnish). Cambridge University Press, Cambridge.

Goebel, K. (1900). *Organography of plants*, Part I, *General organography*, (authorized English edition by I. B. Balfour). Clarendon Press, Oxford.

Goebel, K. (1905). *Organography of plants*, Part II, *Special organography*, (authorized English edition by I. B. Balfour). Clarendon Press, Oxford.

Groenendael, J. M. van (1985). Teratology and metameric plant construction. *New Phytologist*, 99, 171–8.

Groff, P. A. and Kaplan, D. R. (1988). The relation of root systems to shoot systems in vascular plants. *Botanical Review*, 54, 387–422.

Guédès, M. (1966). Sur la valeur du complexe axillaire des Cucurbitacées, II. Organisation et ontogénie des complexes axillaires d'une pousse adulte chez la Bryone (*Bryonia dioica* Jacq). *Bulletin de la Société Botanique de France*, 113, 233–43.

Guédès, M. (1982). A simpler morphological system of tree and shrub architecture. *Phytomorphology*, 32, 1–14.

Hallé, F. (1978). Architectural variation at the specific level in tropical trees. In *Tropical trees as living systems*, Proceedings of the 4th Cabot Symposium, (ed. P. B. Tomlinson and M. H. Zimmermann), pp. 209–21. Cambridge University Press, Cambridge.

Hallé, F. and Delmotte, A. (1973). Croissance et floraison de la Gesnériacée Africaine *Epithema tenue* C. B. Clarke. *Adansonia*, 13, 273–87.

Hallé, F. and Martin, R. (1968). Etude de la croissance rythmique chez l'Hévéa (*Hevea brasiliensis* Müll. Arg. Euphorbiacées - Crotonoidées). *Adansonia N.S.*, 8, 475–503.

Hallé, F. and Ng, F. S. P. (1981). Crown construction in mature Dipterocarp trees. *The Malaysian Forester*, 44, 222–33.

Hallé, F. and Oldeman, R. A. A. (1970). Essai sur l'architecture et la dynamique de croissance des arbres tropicaux. *Collection de monographies de Botanique et de Biologie Végétale*, 6. Masson et Cie, Paris.

Hallé, F., Oldeman, R. A. A. and Tomlinson, P. B. (1978). *Tropical trees and forests: an architectural analysis*. Springer, Berlin.

Hanham, J. (1846). *Natural illustrations of the British grasses*. Binns and Goodwin, Bath.

Harley, J. L. and Smith, S. E. (1983). *Mycorrhizal symbiosis*. Academic Press, London.

Harper, J. L. (1980). Plant demography and ecological theory. *Oikos*, 35, 244–53.

Harper, J. L. (1981). The concept of population in modular organisms. In *Theoretical ecology: principles and applications*, (ed. R. M. May), pp. 53–77. Blackwell, Oxford.

Harper, J. L. (1985). Modules, branches and the capture of resources. In *Population biology of clonal organisms*, (ed. J. B. C. Jackson L. W. Buss, and R. E. Cook). Yale University Press, New Haven.

Hart, J. W. (1989). *Plant tropisms*. Unwin and Hyman.

Hickey, L. J. (1973). Classification of the architecture of dicotyledonous leaves. *American Journal of Botany*, 60, 17–33.

Holttum, R. E. (1954). *Plant life in Malaya*. Longmans, Green and Co.

Jeannoda-Robinson, V. (1977). Contribution a l'étude de l'architecture des herbes. Thèse, Docteur de Specialité de Sciences Biologiques. Université des Sciences et Techniques du Languedoc, Montpellier, France.

Jenik, J. (1978). Roots and root systems in tropical trees: morphologic and ecologic aspects. In *Tropical trees and living systems*, Proceedings of the 4th Cabot Symposium, (ed. P. B. Tomlinson and M. H. Zimmermann), pp. 323–49. Cambridge University Press, Cambridge.

Jong, K. and Burtt, B. L. (1975). The evolution of morphological novelty exemplified in the growth patterns of some Gesneriaceae. *New Phytologist*, 75, 297–311.

Juniper, B. E., Robins, R. J. and Joel, D. M. (1989). *The carnivorous plants*. Academic Press, London.

Kaplan, D. R. (1973a). The teaching of higher plant morphology in the United States. *Plant Science Bulletin*, 19, 6, 9.

Kaplan, D. R. (1973b). The Monocotyledons: their evolution and comparative biology. VII. The problem of leaf morphology and evolution in the monocotyledons. *Quarterly Review of Biology*, 48, 437–57.

Kaplan, D. R. (1975). Comparative developmental evaluation of the morphology of unifacial leaves in the monocotyledons. *Botanische Jahrbücher Syst.*, 95, 1–105.

Kaplan, D. R., Dengler, N. G. and Dengler, R. E. (1982a). The mechanism of plication inception in Palm leaves: problem and developmental morphology. *Canadian Journal of Botany,* 60, 2939–75.

Kaplan, D. R., Dengler, N. G. and Dengler, R. E. (1982b). The mechanism of plication inception in Palm leaves: histogenic observations on the palmate leaf of *Rhapis excelsa. Canadian Journal of Botany,* 60, 2999–3016.

Kirby, E. J. M. (1986). *Cereal development guide,* (2nd edn). Arable Unit, National Agricultural Centre, Stoneleigh, Warwickshire, U.K.

Knuth, P. (1906). *Handbook of flower pollination,* (trans. J. A. Ainsworth-Davies). Clarendon Press, Oxford.

Koriba, K. (1958). On the periodicity of tree-growth in the tropics, with reference to the mode of branching, the leaf-fall, and the formation of the resting bud. *Gardens Bulletin, Singapore,* 17, 11–81.

Krasilnikov, P. K. (1968). On the classification of the root system of trees. In *Methods of productivity studies in root systems and rhizosphere organisms,* (ed. M. S. Ghilarov), pp. 106–114. Nauka, Leningrad.

Kuijt, J. (1969). *The biology of parasitic flowering plants.* University of California Press, Berkeley.

Lloyd, F. E. (1933). The structure and behaviour of *Utricularia purpurea. Canadian Journal of Research,* 8, 234–52.

Mabberley, D. J. (1987). *The plant book: portable dictionary of the higher plants.* Cambridge University Press.

Maddox, G. D., Cook, R. E., Wimberger, P. H. and Gardescu, S. (1989). Clone structure in four *Solidago altissima* (Asteracae) populations: rhizome connections within genotypes. *American Journal of Botany,* 76, 318–26.

Maier, U. and Sattler, R. (1977). The structure of the epiphyllous appendages of *Begonia hispida* var. *cucullifera. Canadian Journal of Botany,* 55, 264–80.

Mallory, T. E., Chang, S.-H., Cutter, E. G. and Gifford, E. M., Jr. (1970). Sequence and pattern of lateral root formation in five selected species. *American Journal of Botany* 57, 800–9.

Mann, L. K. (1960). Bulb organisation in *Allium:* some species of the section Mollium. *American Journal of Botany,* 47, 765–71.

Massart, J. (1921). *Eléments de biologie générale et de botanique.* Maurice Lamertin, Bruxelles.

Mauney, J. R. and Ball, E. (1959). The axillary buds of *Gossypium. Bulletin of the Torrey Botanical Club,* 86, 236–44.

McClure, F. A. (1966). *The bamboos: a fresh perspective.* Harvard University Press, Cambridge, Mass.

Millington, W. F. (1966). The tendril of *Parthenocissus inserta*: determination and development. *American Journal of Botany,* 53, 74–81.

Ming, A., Westphal, E. and Sattler, R. (1988). Proliférations épiphylles provoquées par l'acarien *Eriophyes cladophthirus* chez le *Solanum lycopersicum* et le *Nicandra physaloides* (Solanaceae). *Canadian Journal of Botany,* 66, 1974–85.

Miquel, S. (1987). Morphologie fonctionnelle de plantules d'espèces forestières du Gabon. *Bulletin du Muséum National d'Histoire Naturelle,* Section B, Adansonia. Botanique 9, Section B, No. 1, 101–21.

Mouli, C. (1970). Mutagen-induced dichotomous branching in maize. *Journal of Heredity.* 66, 150.

Mueller, R. J. (1988). Shoot tip abortion and sympodial branch reorientation in *Brownea ariza* (Leguminoseae). *American Journal of Botany,* 75, 391–400.

Ng, F. S. P. (1986). Tropical sapwood trees. In *L'arbre. Compte-rendu du colloque international l'arbre,* pp. 61–7. Naturalia Monspeliensia, numéro hors série.

Noble, J. C., Bell, A. D. and Harper, J. L. (1979). The population biology of plants with clonal growth. I. The morphology and structural demography of *Carex arenaria. Journal of Ecology.,* 67, 983–1008.

Novoplansky, A., Cohen, D., and Sachs, T. (1990). How *Portulaca* seedlings avoid their neighbours. *Oecologia,* 82, 490–93.

Oldeman, R. A. A. (1974). L'architecture de la forêt guyanaise. *Memoires O.R.S.T.O.M.,* 73. O.R.S.T.O.M., Paris.

Pate, J. S. and Dixon, K. W. (1982). *Tuberous, cormous and bulbous plants. Biology of an adaptive strategy in Western Australia.* University of Western Australia Press.

Pijl, L. van der (1969). *Principles of dispersal in higher plants.* Springer, Berlin.

Prévost, M. F. (1967). Architecture de quelques Apocyancées ligneuses. *Mémoires de la Société Botanique de France.* Colloque sur la physiologie de l'arbre, pp. 23–36.

Proctor, M. C. F. and Yeo, P. (1973). *The pollination of flowers.* Collins, London.

Prusinkiewicz, P., Lindenmayer, A. and Hanan, J. (1988). Developmental models of herbaceous plants for computer imagery purposes. *Computer Graphics,* 22, 141–50.

Radford, A. E., Dickinson, W. C., Massey, J. R. and Bell, C. R. (1974). *Vascular plant systematics.* Harper and Row, New York.

Raunkiaer, C. (1934). *The life forms of plants and statistical plant geography,* being the collected papers of C. Raunkiaer (trans. H. G. Carter, A. G. Tansley, and Miss Fansboll). Clarendon Press, Oxford.

Ray, T. S. (1987a). Leaf types in the Araceae. *American Journal of Botany,* 74, 1359–72.

Ray, T. S.(1987b). Diversity of shoot organization in the Araceae. *American Journal of Botany,* 74, 1373–87.

Ray, T. S. (1988). Survey of shoot organization in the Araceae. *American Journal of Botany,* 75, 56–84.

Rees, A. R. (1972). *The growth of bulbs. Applied aspects of the physiology of ornamental bulbous crop plants.* Academic Press, London.

Reffye, P. de, Edelin, C., Françon, J., Jaeger, M. and Puech, C. (1988). Plant models faithful to botanical structure and development. *Computer Graphics,* 22, 151–8.

Reffye, P. de, Edelin, C. and Jaeger, M. (1989). La modélisation de la croissance des plantes. *La Recherche,* 20, 158–68.

Rutishauser, R. (1984). Leaf whorls, stipules and colleters in Rubieae (Rubiaceae) in comparison with other angiosperms. *Beitr. Biol. Pflanz.,* 59, 375–424.

Sachs, J. (1874). *Traité de botanique conforme à l'état présent de la science.* Trad sur la 3° ed par. Van Tieghem, Savy, Paris.

Sattler, R. (1974). A new concept of the shoot of higher plants. *Journal of Theoretical Biology,* 47, 367–82.

Sattler, R. (1982). Proceedings, Developmental Section, International Botanical Congress, Sydney, Australia 1981. *Acta Biotheoretica,* 31A. Martinus Nijhoff/Dr. W. Junk Publishers, The Hague.

Sattler, R. (1984). Homology—a continuing challenge. *Systematic Botany,* 9, 382–94.

Sattler, R. (1986). *Biophilosophy: analytic and holistic perspectives.* Springer, Berlin.

Sattler, R. (1988). Homeosis in plants. *American Journal of Botany,* 75, 1606–17.

Sattler, R., Luckert, D. and Rutishauser, R. (1988). Symmetry in plants: phyllode and stipule development in *Acacia longipedunculata. Canadian Journal of Botany,* 66, 1270–84.

Schellner, R. A., Newall, S. J. and Solbrig, O. T. (1982). Studies on the population biology of the genus *Viola.* IV. Spatial pattern of ramets and seedlings in three stoloniferous species. *Journal of Ecology,* 70, 273–90.

Schmid, R. (1988). Reproductive versus extra-reproductive nectaries: historical perspective and terminological recommendations. *Botanical Review,* 54, 179–232.

Schnell, R. (1967). Etudes sur l'anatomie et la morphologie des Podostémacées. *Candollea,* 22, 157–225.

Schoute, J. C. (1935). On corolla aestivation and phyllotaxis of floral phyllomes. *Verhandeling der Koninklijke akademie van wetenschappene Amsterdam Afdeeling Natuurkunde (Tweede Sectie) Deel XXXIV,* No. 4, 1–77.

Sculthorpe, C. D. (1967). *The biology of vascular aquatic plants.* Edward Arnold, London.

Shah, J. J. and Dave, Y. S. (1970). Morpho-histogenic studies on tendrils of Vitaceae. *American Journal of Botany,* 57, 363–73.

Shah, J. J. and Dave, Y. S. (1971). Morpho-histogenic studies on tendrils of *Passiflora*. *Annals of Botany*, **35**, 627–35.

Shinozaki, K., Yoda, K., Hozumi, K. and Kira, T. (1964). A quantitative analysis of plant form—the pipe model theory. I. Basic analysis. *Japanese Journal of Ecology*, **1**, 97–105.

Silvertown, J. and Gordon, D. (1989). A framework for plant behaviour. *Annual Review of Ecology and Systematics*, **20**, 349–66.

Sporne, K. R. (1970). *The morphology of Pteridophytes*. Hutchinson University Library, London.

Sporne, K. R. (1971). *The morphology of Gymnosperms*. Hutchinson University Library, London.

Sporne, K. R. (1974). *The morphology of Angiosperms*. Hutchinson University Library, London.

Steingraeber, D. A. and Fisher, J. B. (1986). Indeterminate growth of leaves in *Guarea* (Meliaceae): a twig analogue. *American Journal of Botany*, **73**, 852–63.

Stevens, P. S. (1974). *Patterns in nature*. Atlantic Monthly Press Book. Little, Brown and Co., Boston.

Stone, B. C. (1975). Authorized translation of 'An essay on the architecture and dynamics of growth of tropical trees', (F. Hallé and R. A. A. Oldeman). Penerbit University, Malaya, Kuala Lumpur.

Strahler, A. N. (1964). Quantitative analysis of watershed geomorphology. *Transactions of the American Geophysical Union*, **38**, 913–20.

Sutherland, W. J. and Stillman, R. A. (1988). The foraging tactics of plants. *Oikos*, **52**, 239–44.

Taylor, R. L. (1967). The foliar embryos of *Malaxis paludosa*. *Canadian Journal of Botany*, **45**, 1553–6.

Thuret, M. G. (1878). Etudes phycologiques. Analyses d'algues marines, (ed. G. Masson). Librairie de l'academie de médecine, Paris.

Tilney-Bassett, R. A. E. (1986). *Plant chimeras*. Edward Arnold, London.

Titman, P. W. and Wetmore, R. H. (1955). The growth of long and short shoots in *Cercidiphyllum*. *American Journal of Botany*, **42**, 364–72.

Tomlinson, P. B. (1961). Morphological and anatomical characteristics of the Marantaceae. *Journal of the Linnean Society (Botany)*, **58**, 55–78.

Tomlinson, P. B. (1974). Vegetative morphology and meristem dependence—the foundation of productivity in seagrasses. *Aquaculture*, **4**, 107–30.

Tomlinson, P. B. (1978). Some qualitative and quantitative aspects of New Zealand divaricating shrubs. *New Zealand Journal of Botany*, **16**, 299–309.

Tomlinson, P. B. (1983). Tree architecture. *American Scientist*, **71**, 141–9.

Tomlinson, P. B. (1984). Homology: an empirical view. *Systematic Botany*, **9**, 374–81.

Tomlinson, P. B. (1987). Branching is a process not a concept. *Taxon*, **36**, 54–57.

Tomlinson, P. B. (1990). *The structural biology of palms*. Oxford University Press.

Tomlinson, P. B. and Bailey, G. W. (1972). Vegetative branching in *Thalassia testudinum* (Hydrocharitaceae)—a correction. *Botanical Gazette*, **133**, 43–50.

Tomlinson, P. B. and Esler, A. E. (1973). Establishment growth in woody monocotyledons native to New Zealand. *New Zealand Journal of Botany*, **11**, 627–44.

Troll, W. (1935–43). *Vergleichende morphologie der höheren pflanzen*. Published progressively. Verlag von Gebrüder Borntraeger, Berlin.

Troll, W. (1964). *Die infloreszenzen. Typologie und stellung im aufbau des vegetationskörpers*. Teil 1, Jena.

Tucker, S. C. and Hoefert, L. L. (1968). Ontogeny of the tendril in *Vitis vinifera*. *American Journal of Botany*, **55**, 1110–9.

Tulasne, L. R. (1852). *Podostemacearum monographia*. Archives du Museum d'histoire naturelle, tome VI, Paris.

Velenovsky, J. (1907). *Vergleichende morphologie der pflanzen, teil II*. Verlagsbuchhandlung von Fr. Rivnac. Druck von Eduard Leschinger, Prag.

Vincent, J. R. and Tomlinson, P. B. (1983). Architecture and phyllotaxis of *Anisophyllea disticha* (Rhizophoraceae). *Gardens Bulletin Singapore*, **36**, 3–18.

Ward, H. M. (1909). *Trees*. Volume V. *Form and habit*. Cambridge University Press. Cambridge.

Weberling, F. (1965). Typology of inflorescences. *Journal of the Linnean Society (Botany)*, **59**, 215–21.

White, J. (1984). Plant metamerism. In *Perspectives on plant population ecology*, (ed. R. Dirzo and J. Sarukhán), pp. 15–47. Sinauer Associates, Massachusetts.

Willis, J. C. (1960). *A dictionary of the flowering plants and ferns*, (6th edn.). Cambridge University Press, Cambridge.

Willis, J. C. (1973). *A dictionary of the flowering plants and ferns*, (8th edn.). Cambridge University Press, Cambridge.

Young, J. P. W. (1983). Pea leaf morphogenesis: a simple model. *Annals of Botany*, **52**, 311–6.

Zimmermann, M. H. and Brown, C. L. (1971). *Trees: structure and function*. Springer, New York.

Index